Evolution
The General Theory

ERVIN LASZLO

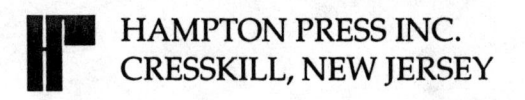

HAMPTON PRESS INC.
CRESSKILL, NEW JERSEY

Copyright © 1996 by Hampton Press, Inc.

All rights reserved. No part of this publication may be reproduced, storied in a retrieval system, or transmitted in any form or by any means, electronic, mechanical, photocopying, microfilming, recording, or otherwise, without permission of the publisher.

Printed in the United States of America

Library of Congress Cataloging-in-Publication Data

Laszlo, Ervin, 1932-
    Evolution : the grand synthesis / Ervin Laszlo
        p. cm. -- (Advances in systems theory, complexity, and the human sciences)
    Includes bibliographical references and index.
    ISBN: 1-57273-052-8
    1. Evolution (Biology) 2. Social evolution. I. Title
II. Series.
QH371.L37 1996
575--dc20   96-9510
                                                                                    CIP

Hampton Press, Inc.
23 Broadway
Cresskill, NJ 07626

Contents

Introduction 1

PART 1: THE LOGIC OF EVOLUTION

1. The Rise of the Evolutionary Paradigm 11
    Antecedents of the New Paradigm 12
    The Puzzle of Two Arrows of Time 16
2. Outline of the General Theory 21
    Basic Concepts 22
        Equilibrium, Determinism, and Scientific Law ... 22
        Systems in the Third State 23
        The Continuum of Evolution 25
    Empirical Findings 30
        Structure and Free Energy 30
        Convergence to Higher Levels 34
        Instability and Transformation 37
    Theoretical Findings 40
        Autopoiesis 40
        Catastrophes, Chaos, and Bifurcations 42
    Summary 46
    Notes 53

Contents

PART 2: THE REALMS OF EVOLUTION

Introduction	59
3. The Evolution of Matter	63
The Origins of the Universe	63
Grand Unified Theories (GUTS)	64
Inflationary Scenarios	65
The Synthesis of Matter-Energy Systems	67
Temperature and Density Parameters	67
The Arrow of Time in the Cosmos	72
The Radiation-to-Matter Energy Transfer	72
Notes	75
4. The Evolution of Life	79
The Origins of Biologic Evolution	80
The Dynamics of Biologic Evolution	83
Speciation	83
Mutation	87
Convergence	89
The Emergence of Sapiens	91
Notes	94
5. The Evolution of Society	95
The Evolutionary Axioms of Society	97
The Direction of Social Evolution	99
Technology as Engine of Change	100
The Dynamics of Social Evolution	108
Bifurcations in History	108
The Varieties of Social Bifurcations: A Handful of Case Studies	114
T-Bifurcations	114
C-Bifurcations	119
E-Bifurcations	123
Conclusions	125
Notes	126
6. The Evolution of Mind	127
The Origins of Mental Faculties	127
The Nature and Continued Evolution of Mind	129
A Concluding Note	135
Notes	135
From the Midnight Notebook	137
Appendix: The Evolution of Science	141
Notes	147
Bibliography	149
Index	155

Introduction

Everybody knows the meaning of "evolution." Children know it as the theory that man descended from the monkeys and not from Adam and Eve. Adults know it as the theory of Darwin, that all living species had a common origin. Biologists know it as the "modern synthesis," the neo-Darwinian integration of biological theories in which mutation and natural selection account for the variation and emergence of species.

All these conceptions are true, as far as they go. None go as far as the meaning of "evolution" in the sense of the currently emerging paradigm of scientific thought. In this sense the concept of evolution goes beyond the origins and development of all living species. It embraces the patterns and dynamics of change in the cosmos as well as in the living world; in the history of human culture and society no less than in the history of life on earth. In its emerging meaning evolution is not only the evolution of living species but the evolution of all things that emerge, persist, and change or decay in the known universe. It is evolution in the generalized sense of the term, and the theory that describes it is GET: general evolution theory.

Now "evolution" is not a newly invented concept—it comes from the Latin *evolvere*, meaning to unfold. It was first applied, erroneously as it turned out, to the development—or "unfolding"—of the full-grown organism from the minute homunculus that was presumed to exist, fully formed, in the male sperm or in the female egg. Later the concept of evolution became identified with the theory of Darwin and the field of macrobiology. Notwithstanding the encompassing evolutionary philosophies of Herbert Spencer, Henri Bergson, Samuel Alexander, Alfred North Whitehead, and Teilhard de Chardin, and the misguided attempts of Social Darwinists to make the struggle for survival in the human sphere into a social and political doctrine of modern society, evolution remained restricted essentially to biological theory—until, that is, a group of new disciplines that came to be collectively known as "sciences of complexity" entered the scene. These disciplines, which include general system theory, cybernetics, information and communication theory, dynamics, autopoietical system theory, as well as catastrophe, chaos, and dynamical system theory and, above all, nonequilibrium thermodynamics, began to describe irreversible processes of change and transformation in a rigorous fashion. Their findings proved to have application to a wide range of phenomena, from physics to chemistry, from biology to ecology, and from historiography to psychology, sociology and the allied social sciences such as organization and management theory and the theory of international relations.

The realization that change is irreversible in nature as well as in certain fields of human and social development and, even more, the recognition that such change exhibits analogous dynamic patterns in domains that are seemingly entirely different, led to a systematic search for commonalities that would underlie its various manifestations. Invariances in the dynamic and formal aspects of complex systems were actively researched in general system theory and cybernetics since mid-century but were centered mainly on processes of self-preservation, operating by means of self-correcting negative feedback. Processes of self-transformation, that is, fundamental and irreversible change, came into the focus of investigation in the 1960s as Prigogine, Katchalsky, Curran, de Groot, Nicolis and others began to publish their path-breaking theories. At about the same time the new cosmology made its appearance in the work of Shapley, Weinberg, Guth,

Hawking and others, and proved to be a fertile field for the exploration of continuities between the evolution of physical structures in the universe and the structures of the living world here on earth. The study of irreversible change was reinforced by new developments, such as the topological theories of Thom and Zeeman, and the chaos theory elaborated by Birkhof, Rössler, Abraham, Shaw, and others on the basis of the pioneering work of Edward Lorenz in meteorology and Benoit Mandelbrot in mathematics. As Peter Allen and the Brussels school began to extend the theories of irreversible thermodynamics to the living and to the social spheres, and Maturana and Varela began to investigate cognitive processes in light of autopoietic system theory, the stage was set for a thorough exploration of the phenomenon of evolution in its full breadth, from cosmos to culture. Chaisson in cosmology, Artigiani and Eisler in history, Csanyi, Salthe, and Corliss in the life sciences, Loye and Schull in psychology, Ceruti and Bocchi in philosophy, Banathy in human development and education theory, and Salk in its general human and social implications, are among the pioneers in this field. My own work over the past years has been devoted to the creation of GET: a general theory that goes beyond the exclusively biology-oriented modern synthesis to join the developmental aspects of cosmology with similar aspects of biology, and the human and social sciences.

GET satisfies a basic ideal of science: to find embracing, encompassing, and coherent meaning underlying the chaotic welter of everyday experience. The search for meaning is not limited to science: it is constant and continuous—all of us engage in it during all our waking hours; the search continues even in our dreams. There are many ways of finding meaning, and there are no absolute boundaries separating them. One can find meaning in poetry as well as in science; in the contemplation of a flower as well as in the grasp of an equation. We can be filled with wonder as we stand under the majestic dome of the night sky and see the myriad lights that twinkle and shine in its seemingly infinite depths. We can also be filled with awe as we behold the meaning of the formulae that define the propagation of light in space, the formation of galaxies, the synthesis of chemical elements, and the relation of energy, mass and velocity in the physical universe. The mystical perception of oneness and the religious intuition of a Divine intelligence are as much a construction of meaning as the postulation of the universal law of gravitation.

The search for meaning takes many forms; many kinds of meanings can be found. It is up to us which ones to accept. Our choice is determined by the criteria for meaning that we choose to adopt. Science does not differ from art and religion in intrinsic meaningfulness, but it does differ in its criteria of acceptability. These criteria are stated in the method of science; it is by adherence to its method that scientists admit or reject concepts and hypotheses. The method of science involves hypotheses that are tested against experience—against direct observation or the reading of instruments—with the proviso that the test be repeatable at all times and by all people in identical circumstances. And if a hypothesis is borne out by experience, it is compared with alternative hypotheses and is accepted only if it explains more with less: if it applies to a wider range of phenomena with fewer assumptions than any other. Einstein said it clearly: "we are seeking for the simplest possible scheme of thought which will bind together the observed facts" [*The World As I See It*, 1934]. Art, religion and non-scientific systems of thought in general, do not have to respond to these particular criteria. But they have their own criteria, their own methods of validation. Not everything that people play on an instrument is great music; not every configuration of paint that they dab on a canvas is great art; not every intuition of a higher reality is religion.

The way science derives meaning from experience is not necessarily better than any other; in some respects it is more limited and hence less satisfactory. But science excels in one respect at least: it is the system of thought that is the most consistent, and the most thoroughly tested and hence reliable. If a hypothesis is not public and testable, it is not science; and if it is not the most parsimonious, the most coherent and embracing of all that are advanced it is not *valid* science.

Given the restrictions on acceptability imposed by the method of science, it is indeed remarkable that it could make enormous progress in constructing the stream of immediate experience into a world that, while often abstract, is nevertheless consistent, minimally burdened with a priori assumptions, and embracing of a growing range of phenomena.

The unity that contemporary science seeks and increasingly finds is not in the form of a fundamental element from which the manifest diversity of the world is built up, and to which it can be reduced. Rather, it is in the form of the fundamental pattern that appears in ever more varied, more diverse transformations. The key concept is not a substance, it is not even the form of a substance. It is not any form or formula of physics or biology, but a pattern that transcends every empirical discipline and embraces them all. To use the classical Greek expressions, the "one" is not the form of *Being* but the form of *Becoming*. In modern terms it is not the form of the brick but the form of the building. And in the terminology of science it is not an element, a cell, or some other "basic" unit but the pattern of irreversible change manifest in all systems far from thermodynamic equilibrium. The key concept is the invariant pattern of evolution.

It is a source of the deepest wonder that this pattern of change, the pattern of evolution, exhibits a basic unity and consistency. After all, this need not have been the case: nature is not constrained to be logical. Yet there is logic in nature, and unity. There is an inherent order that underlies and interrelates all the phenomenal orders that appear in an almost infinite variety. There is, in the pattern traced by nature evolving, an order that repeats, an invariance that is conserved. The ultimate "one" that underlies the experienced "many" is the invariance in the evolution of complexity in physical nature, in the living world, and in the world of humankind.

What, precisely, is the nature of this invariance? It is, I believe, (as does David Bohm) the order of change itself. It is the order of orders; the order that orders exhibit when they emerge in the universe. It is the order of evolution. Evolution is truly an unfolding, but not of things or substances but of orders. The order inherent in the physical universe unfolded first, appearing already at the first 10^{33} second that marked the end of "Planck-time"—and the beginning of the cosmic processes that still hold sway in today's universe. The order that arose some ten or twelve billion years later was the biophysical and biological order exhibited by self-replicating and self-sustaining thermodynamically open systems basking in a rich flow of energy on suitable planetary surfaces. And the order that precipitated from these higher-level orders on our own planet is the order of the human world, the order wrought by thought and feeling and intuition, and expressed in the societies and cultures created by thinking and feeling human beings.

It used to be thought—and it still is thought by "pragmatic" specialists who willingly wear the blindfolds of their specialties—that there is no discernible relation between the manifest orders of the physical, the living, and the human worlds. It is practical and efficient to think so as long as *ad hoc* assumptions work—and they work as long as one is digging near the surface of phenomena where almost anything that one turns up is new and significant. The economist investigating the effect of a change in the money-supply on external trade has no need to worry about the evolution of galaxies, nor about the way in which human societies have evolved from hominid tribes. The biologist at pains to understand the influence of irradiation on the dynamics of phase change in the genome need not concern himself with nuclear processes in the interior of stars. But the mind that seeks order in experience is not halted at the boundaries of disciplinary specialties. It cannot help asking if there is not some connection between these seemingly diverse processes. Is there not an order that connects the emergence of living systems on the surface of certain planets with the formation of the planets themselves? Is there not an order that relates the emergence of societies populated by self-interest driven economic actors with the mutation of genomes in organisms?

These and similar questions seem arcane; any answer we might give to them would appear far-fetched. Yet the remarkable, indeed the momentous, fact in the development of the contemporary sciences is that such questions not only can be answered, but that the answer we can give to them is coherent and unitary. Evolution, we now see, repeats itself. It is not that it is the same in the different domains, but that its basic dynamic and formative features are invariant. The basic descriptions that we can now give of the processes of evolution remain unchanged as we move from the physical to the biological, and from the biological to the sociocultural realms. There are general laws of evolution, and these general laws refer to invariant patterns appearing in diverse transformations. They are the warp and woof of GET: the general theory of evolution.

The recognition that, underneath the great diversity of empirical phenomena, there is a fundamental invariance, an order that governs the unfolding of order in the universe inspires the same depth of awe and wonder as great art, and great religious or mystical experience and intuition. At long last we may be coming face to face with the reality that the human mind has perennially sought and occasionally glimpsed, but never truly grasped. We may now be realizing the inspiration of the artist and the intuition of the mystic as we realize the ambition of the scientist. We may come closer than ever before to beholding the "sublimity and marvelous order which reveal themselves both in nature and in the world of thought"—to quote Albert Einstein again.

Introduction

The search for this order has motivated the efforts of great minds throughout the history of human consciousness. That it now brings fruit within the rigorous limits of the scientific method should be a cause for joy and encouragement in an age when science is more feared than revered, and more noted for creating technologies that destroy than systems of thought that enlighten.

Part 1

The Logic of Evolution

1

The Rise of the Evolutionary Paradigm

Systems scientific and philosophic, said Alfred North Whitehead, come and go. Each method of limited understanding is at length exhausted. In its prime each system is a triumphant success: in its decay it is an obstructive nuisance. Whitehead's dictum is true of the systems and methods of the contemporary sciences. The triumphant success of the Newtonian system became an obstructive nuisance at the beginning of the century when Einstein's theory fought for acceptance, Darwin's synthetic theory remained an obstruction in the way of the newer, more adequate theories of species evolution in the 1970s, and the positivistic conception of history remains an obstacle in the way of sustained attempts to inquire into patterns in the evolution of human societies to this day.

But a new system, scientific in origin and philosophic in depth and scope, is now on the rise. It goes beyond the limited understanding conveyed by theories within delimited fields of inquiry. It encompasses the great realms of the material universe, of the world of the living, and of the world of history. This is the evolutionary paradigm: the framework for studying the adventure of open systems as they pursue their careers in spheres of nature ever further removed from the inert state of thermodynamic equilibrium, ever more complex and dynamic, ever more in control of themselves and of their environment.

The advent of the new paradigm is a major step in the advance of contemporary science. Science may not be provable, but, as Sir Karl Popper said, it is improvable. The new evolutionary paradigm is an improvement over the previous, discipline-bound theories and paradigms, for it explains more with less. Its basic assumptions and fundamental theorems remain unchanged as we pass from the realm of matter to the sphere of life, and from the sphere of life to the domain of history. Its simplicity and scope are consistent with enduring values that define the course of scientific progress. Simplicity, "elegance," extensibility, consistency, and scope of applicability have always motivated the efforts of scientists; they came clearly to be expressed in the works of Copernicus and Newton, Einstein and Heisenberg. Today they find expression in what has sometimes been called the "evolutionary vision"—the basic set of concepts that defines the interdisciplinary paradigm of system change, persistence, and transformation in the world at large.

The evolutionary paradigm is emerging in field after field; its validity is acknowledged by more and more scientists. They recognize that the new paradigm can suggest concrete ways and means to solve many of the puzzles in their own special yet subtly interconnected discipline. The new paradigm marks the coming of a new era in scientific thinking: an era in which evolution, expressed in human beings and in human societies, is becoming conscious of itself.

Antecedents of the New Paradigm

The new paradigm of evolution did not emerge, like Venus from the sea, all at once, fully created and perfect in every detail. Fundamentally similar ideas have inspired many of the great intellectual currents of human civilization, from Taoism in the East to the Ionian natural philosophies in the West. But these ideas were based on speculation and insight rather than scientific data and tested hypothesis. It is only in our own day that the evolutionary paradigm could reappear within the more tested, and hence more reliable, framework of science itself.

Without pretending to offer anything like an exhaustive historical account, we can sample some features of Western intellectual history to trace the fortunes of synthesizing evolutionary thinking and its recent entry within science proper. We begin with the ancient Greeks, the first to examine the myths of creation in the cool light of reason.

Almost three thousand years ago, the Ionian natural philosophers began to create a grand evolutionary synthesis that would explain the origin and nature of all things through basic and mutually consistent concepts. They were struck by the diversity of the things encountered in experience and applied themselves to showing how it could have come about from common origins.

In the sixth century before Christ, Thales taught that all things in the world originate from a common source, which he identified as water. His contemporary Anaximander did not regard water as the sole substance from which all others arise. He did not identify the precise nature of the original substance although he did say that it was unfamiliar and limitless and encompassed all the worlds. Earth, fire and water were constantly transforming according to an ordering principle based on justice. Anaximenes, a student of Anaximander, speculated that the primeval substance was a mixture of water and earth. Warmed by the sun, the mixture generated plants, animals, and human beings by spontaneous creation. Air, being of even greater importance than either water or earth, produced all substances through condensation. Fire was but rarefied air that, when condensed, became cloud, water, earth, and ultimately stone.

Among the classical Greeks the wish to be consistent led to further syntheses, of a more or less evolutionary character. Heraclites, who thought fire the most important of the substances, placed stress on eternal becoming: on the principle that "change is all." One cannot step into the same river twice; one cannot identify anything in the world for what it truly is; it is constantly changing. Empedocles, in turn, viewed all things as composed of air, earth, fire, and water in measures determined by the principle of love, which binds, and of hate, which separates. From fires within the earth arose the primal forms that later evolved into the familiar organisms. Many were imperfect and disappeared, while those that proved perfect survived.

With Socrates the naturalism of the pre-Socratics became tuned to the human world: man is the measure. Plato maintained a systematic view of the whole of reality, if only as a "likely tale": he believed all things in the perceived world to be images or shadows of eternal and unchanging Forms or Ideas. Aristotle replaced this conception with a new, more careful naturalism based in part on observation integrated in a truly encyclopedic range of knowledge. His "great chain of being" extended from inanimate objects through plants and animals to humans. The progressive development of nature was matched by a ripening of the soul. The inorganic became the organic through metamorphosis, and in the realm of the organic, animals with powers of sensibility were more animate than plants, endowed merely with the powers of nourishment and sensibility. Nature proceeded gradually and constantly from the least to the most perfect, becoming ever more complex in the process. Nothing, Aristotle said, exists without a cause; the progression is not accidental: it is motivated by a final cause, and this is perfection.

Without revering the Greek philosophers as much as Whitehead, who claimed that all Western philosophy is but a series of footnotes to Plato, or following in the footsteps of the scholastics, whose aim was mainly to reconcile the Christian doctrine with the teachings of Aristotle, we can nevertheless skip over the intellectual history of the period between the late Hellenic and the modern Renaissance: although it includes Epicurus and Lucretius as well as Augustine and Thomas Aquinas, it did not contribute much to the advancement of an embracing evolutionary paradigm. A new beginning was made only with the emancipation of science from church dogma in the Renaissance of the fifteenth century. This vas truly a beginning, a restatement of the most elementary observations in a language devoid of reliance on either Aristotle or on theology. But with exceptional geniuses like Leonardo even minute observations ranged over enormous terrain, from marine fossils to shadows on the moon. Copernicus, however, was already a scientific specialist, though with a deep belief in metaphysical principles such as "Nature loves simplicity." His quest to understand celestial mechanics through the simplest possible system of explanation led to the heliocentric hypothesis, reviving and substantiating an idea already known to the ancient Greeks. Galileo, experimenting with inclined planes and falling objects, advanced the concept of a law of gravitation, while Kepler, equipped with the telescope and with an analytical mind, removed the vestiges of metaphysical rationalism from Copernican theories by showing that the planets described ellipses, and not perfect (and perfectly simple) circles around the sun. Newton, producing perhaps the greatest synthesis of all times, laid the foundations for modern physics by uniting the terrestrial with the celestial sphere through the universal law of gravitation and the equally universal laws of motion.

The emancipation of science led to stress on experiment instead of dogma, and experiment, given the primitive means and instruments available, privileged problems that could be solved with basic formulas of universal application. With Newtonian physics the physical universe became a uniform mechanism which, once wound up, churned out harmonious motions for all eternity. The world view of modern science became mechanistic and soulless, in full contradiction with the world of the living. With the splitting of the known world into a mechanistic universe and a separate realm of life and mind, attempts to create an embracing science of evolution and development encountered an obstacle as formidable as Christian theology. Natural philosophy became divorced from moral philosophy, physics from biology, and the natural sciences from the human and social disciplines.

With Descartes the separation of the realms of matter and mind achieved the status of metaphysical doctrine. The physical universe, including the bodies of living organisms, was but a mechanism, describable by a small number of God-given physical laws. Human mind and consciousness were removed from this sphere and became a separate substance, existing not in space but only in time. Seeking to overcome dualism in his naturalist ethics, Spinoza embraced pantheism, identifying nature with God. Kant, who in his early work on universal natural history asserted that all observable phenomena have natural causes, and who attempted to trace the development of higher forms of life to elementary designs, succumbed in his influential *Critiques* to the posture that separated the knower from the known, and made the natural world into an unattainable "noumenon."

Not until the puzzle of the two and opposing arrows of time was resolved in the late twentieth century was there a sound basis for bridging the gap between matter and mind, the natural world and the human world, and thus between the "two cultures" of modern Western civilization.

The Puzzle of Two Arrows of Time

Although by the early eighth century scientists tended to view the world as the result of changes that took place in its past, science was stymied even in the nineteenth century by the persistent contradiction between a mechanistic world slated to run down and an organic world seeming to wind up. With respect to the latter, creationism proved to be a temporary expedient with its claim that species did not evolve (and hence the living world did not really progress in a direction contrary to physical nature)—they were special creations of a supreme mind. Linnaeus, who founded the modern system of classification in botany and biology, said that "there are as many species as originally created by the Infinite End." Organisms, in the view of many biologists, did not reproduce: they "engendered." Generation was the result of the creative act which logically called for God, the great craftsman. Physicists themselves could say nothing about life and its origins and evolution: Newton bracketed out the irreversible changes that lie at the root of all evolutionary processes. The laws of classical mechanics apply regardless of whether a process or particle proceeds from point A to point B or from point B to point A. As particles move in time as well as in space, in Newtonian physics both time and motion are fully reversible.

By the middle of the nineteenth century the physical and the life sciences found themselves enmeshed in endless contradictions concerning the nature of change. Among the physical sciences classical thermodynamics dealt with one-way processes in nature and admitted the irreversibility of time. In the life sciences Darwin's theories of the origin of species, based on one-way processes of cumulative change in the world of the living, created a veritable revolution. The evolutionary processes postulated in classical thermodynamics and Darwinian biology conflicted with classical physics; they also conflicted with each other. The one-way processes discovered by classical thermodynamics and Darwinian biology did not match. In classical thermodynamics the "arrow of time" points downward, toward states of disorganization and randomness, while in Darwinian biology it points upward, toward higher levels of organization in determinate structures and functions. Since Newton's classical mechanics was not challenged until the beginning of the present century, nineteenth century natural science found itself saddled with two arrows of time—and a framework of classical physics that knew nothing of either.

The imagination of philosophers and social scientists was captured first and foremost by the upward sweep of time in the Darwinian theory. Building on the concepts of German biologists and greatly encouraged by the publication of *The Origin of Species*, the English philosopher Herbert Spencer published his *First Principles*—a comprehensive, science-based account of evolution as leading to the greatest perfection and the most perfect happiness. Likewise impressed by Darwin, Marx and Engels followed the upward sweep of Hegel's evolutionary optimism in their encompassing theory of evolution in nature and in society, known as dialectical and historical materialism.

These were among the noteworthy attempts of nineteenth-century philosophers and scientists to formulate a grand evolutionary theory based on the findings of modern science. But in today's hindsight, their efforts were premature. The puzzle of the two arrows of time could not be resolved within nineteenth-century science, and general theories based on such science had to opt between the downward-pointing arrow of thermodynamics and the upward-oriented arrow of biology. Spencer and Marx, among others, opted for the upward sweep of Darwin and ignored the persistent degrading effect of time in classical thermodynamics.

Classical thermodynamics, however, could not be ignored for long. Its basic concepts were formulated by Sadi Carnot in 1824 and elaborated by William Thompson in his 1852 treatise "on the universal tendency in nature to the dissipation of mechanical energy." In 1865 Rudolf Clausius introduced the quantitative definition of entropy—the measure of organization versus randomness in a system—and in 1872 Boltzman advanced the formulation that linked thermodynamics with statistical mechanics and established the discipline as a major field in physical science.

The famous Second Law of Thermodynamics affirmed that for any given system conceived as a closed—in fact, an isolated—unit of matter and energy, differences and gradients in concentration and temperature tend to disappear, to be replaced by uniformity and randomness. The universe, at least its material components, moves from a more organized and energetic state toward states of growing homogeneity and randomness. It ultimately reaches the state of perfect heat distribution in which no irreversible processes can any longer occur: there are no hotter and colder bodies to create a flow of energy. The arrow of time is thus given by the probability that closed systems run down, that they move *toward* equilibrium rather than *away* from it.

Although some nineteenth-century philosophers and social scientists ignored it, the conflict between physics and the life sciences became more and more evident. Some of the greatest minds of the epoch began to be preoccupied by it. Of the many ingenious suggestions for resolving it, two deserve special mention. One was made by the French philosopher Henri Bergson. The universe, said Bergson, shows two tendencies: there is a "reality which is making itself in a reality which is unmaking itself." The general tendency to repetition and the dissipation of energy, discovered by thermodynamics, is the tendency of "matter"; the countertendency within it, documented by Darwin, is the tendency of "life." The evolution of life is the result of the workings of a basic impulse called the *élan vital*. It thrusts itself forward into ever-new forms of organized structure. These structures store and utilize energy and maintain their capacity for growth and adaptation up to a point; then they lapse into repetitive routine and ultimately degrade their vital energy.

Another noteworthy suggestion was put forward by Ludwig Boltzman, one of the founders of classical thermodynamics. Conceive of the current state of the universe, he said, as but a brief deviation from equilibrium. Our world is only one of several regions which are in disequilibrium; among these worlds the probabilities of their state (i.e., their entropy) increase as often as they decrease. In the universe as a whole the two directions of time are indistinguishable, so that a living organism in one of these regions can always define the direction of time as going from the less probable to the more probable state. The former state is its "past" and the latter its "future."

However ingenious, Boltzman's proposal could no more resolve the conflict between the Second Law of Thermodynamics and Darwinian evolution than that of Bergson. Postulating two layers of reality, one called matter and the other life (the latter infused by a mysterious *élan vital*), and making our world into a brief interval balanced by others tending in the opposite direction (with all such locally disbalanced regions having arbitrary definitions of time), makes for speculative cosmology rather than sound science. Scientists had to wait for nonequilibrium thermodynamics and the new cosmology of the 1970s to perceive that there is no need for metaphysical solutions to the conflict between upward-tending biological evolution and the downward-tending processes of closed physical systems.

We have now come to see that the conflict between the two great processes—the two arrows of time—is only apparent.

Evolving systems are not closed; the universe as a whole is not mechanistic; cosmic processes do not point the arrow of time toward a state of universal heat-death; and life is neither an accidental aberration nor the manifestation of mysterious metaphysical forces.

It took the better part of two millennia for the evolutionary paradigm to reassert itself within the more solid framework of empirical science, and almost two centuries for empirical science to emancipate itself from the dogmas of the Middle Ages. But in our day at last, the evolutionary paradigm surfaces as the dominant mode of thought underlying work in almost all of the natural sciences, and in many of the social sciences as well. Not withstanding remnants of compartmentalization and pockets of resistance (by scientific specialists as well as dogmatic creationists), evolutionary thinking is here to stay. If integrated within a general yet sound theory, its results can add up to a new and more fruitful conception of the nature of reality, in the universe at large as well as here on earth. Those who make creative use of the new theory will be the leading minds of our time, whose vision will open new vistas for science and fresh opportunities for people and societies.

2

Outline of the General Theory

In the penultimate decade of the twentieth century science is sufficiently advanced to resolve the puzzles that stymied scientists in the last century and demonstrate, without metaphysical speculation, the consistency of evolution in all realms of experience. It is now possible to advance a general evolution theory based on unitary and mutually consistent concepts derived from the empirical sciences. We can lay the foundation for this major interdisciplinary endeavor by assembling and analyzing the basic concepts of the contemporary sciences of evolution.

Today, the processes and manifestations of evolution are investigated not only in contemporary variants of classical disciplines, but also in a handful of freshly emerging fields. The new sciences deal with the appearance, development, and functioning of complex systems regardless of the domain of investigation to which they belong. They originated with the general system theory pioneered by Ludwig von Bertalanffy, Paul Weiss, Anatol

Rapoport, and Kenneth Boulding, and with the sciences of cybernetics developed by Norbert Wiener, W. Ross Ashby, and Stafford Beer. Since the 1960s they were joined and reinforced by nonequilibrium thermodynamics, the work of Aharon Katchalsky and Ilya Prigogine and their followers, by cellular automata theory pioneered by John von Neumann and evolved in the autopoietic system theory of Humberto Maturana and Francisco Varela, and by catastrophe theory and dynamic systems theory developed by René Thom, Christopher Zeeman, Robert Shaw, and Ralph Abraham, among others. As these fields of science—known collectively as sciences of complexity—offer the most logical foundation for GET, we now review their pertinent concepts and conceptions.

BASIC CONCEPTS

Equilibrium, Determinism, and Scientific Law

We can best introduce the fundamental concepts of the new evolutionary sciences by contrasting their basic features with some received notions of more traditional theories.

The contrast is indeed striking. Traditionally the natural sciences viewed evolution as a basically deterministic process oriented toward equilibrium. Factors of chance and instability were under-emphasized in favor of stability, control, and predictability. While in physics randomness was conceptualized in quantum mechanics, it was limited to the realm of the microscopic; at macroscopic levels phenomena were considered to be basically deterministic. The mechanistic world view underlying Newtonian physics, and the Laplacean belief that, given enough information, we could predict the position of every particle in the universe at any time, was highly persuasive and long dominant. The laws of nature had to be deterministic as well as universal; exceptions to them were viewed with mistrust and ascribed to flaws in the system of reckoning. The states toward which systems move in many complex but basically deterministic processes were thought to be states of equilibrium—of relaxed tensions, rest and balance.

Concepts of equilibrium and determinacy were dominant also in the sciences of life and society. Biologists resorted to the deterministic logic whereby natural selection brings about a proper balance between predator and prey and assures the survival of the fittest under predictable conditions. Economists, following in the footsteps of Adam Smith, are still loath to surrender the central place of equilibrium in their theories (Kenneth Boulding may have been right when he remarked of modern economics that "nothing fails like success"): general equilibrium theory is not, as yet, an expression of the evolutionary paradigm. Neither is the mainstream theory of the social scientists, centered as it is on statics instead of dynamics; structures and states instead of processes and functions; self-correcting mechanisms instead of self-organizing systems; and conditions of equilibrium instead of dynamic balances in regions of distinct nonequilibrium.

The evolutionary paradigm challenges concepts of equilibrium and determinacy in scientific theories; and it modifies the classical deterministic conception of scientific laws. The laws conceptualized in the evolutionary context are not deterministic and prescriptive: they do not uniquely determine the course of evolution. Rather, they state ensembles of possibilities within which evolutionary processes can unfold. They are the rules of the game, to be exploited in each individual instance according to the skills and predispositions of the players. Dynamic systems—the "players" in the game of evolution—are not strictly determined. They do not have individual trajectories (evolutionary paths) but bundles of trajectories. This property undermines classical determinism, based on the concept of a single trajectory. Given identical initial conditions, different sequences of events unfold—within the limits and possibilities set by the laws. The sequences in turn create fresh sets of limits and possibilities. Evolution is always possibility and never destiny. Its course is logical and comprehensible, but it is not predetermined and thus not predictable.

Systems in the Third State

A general theory of the myriad forms and processes of evolution is possible because, as the sciences of complexity now discover, evolution is nevertheless singularly consistent: it brings forth the same basic kind of entity in all its domains. The entity is a system of a particular kind—a system that we may best identify as "system in the third state."

The fact is that systems in the real world can exist in one of three types of states. Of these three, one is radically different from classical conceptions: it is the state far from thermal and chemical equilibrium. The other two states are those in which systems are either *in* equilibrium or *near* it. In a state of equilibrium, energy and matter flows have eliminated differences in temperature and concentration; the elements of the system are unordered in a random mix, and the system itself is homogeneous and dynamically inert. The second state differs only slightly from the first: in systems near equilibrium there are small differences in temperature and concentration; the internal structure is not random and the system is not inert. Such systems will tend to move toward equilibrium as soon as the constraints that keep them in nonequilibrium are removed. For systems of this kind, equilibrium remains the "attractor" which it reaches when the forward and reverse reactions compensate one another statistically, so that there is no longer any overall variation in the concentrations (a result known as the law of mass action, or Guldberg and Waage's law). The elimination of differences between concentrations corresponds to chemical equilibrium, just as uniformity of temperature corresponds to thermal equilibrium. While in a state of nonequilibrium the system performs work and therefore produces entropy, at equilibrium no further work is performed and entropy production ceases.[1]

The third possible state of systems is the state far from thermal and chemical equilibrium. Systems in this state are nonlinear and occasionally indeterminate. They do not tend toward minimum free energy and maximum entropy but may amplify certain fluctuations and evolve toward a new dynamic regime that is radically different from stationary states at or near equilibrium.

Prima facie, the evolution of systems in the third state appears to contradict the Second Law of Thermodynamics. How can systems actually increase their level of complexity and organization, and become more energetic? The Second Law entails that in any isolated system organization and structure tend to disappear, to be replaced by uniformity and randomness. But unlike scientists in the nineteenth century, contemporary investigators realize that evolving systems are not isolated systems, and thus the Second Law does not fully describe what takes place in them—more precisely, between them and their environment. The fact we now realize is that systems in the third state are open systems, and the change of entropy within them is not determined uniquely by irreversible processes within their boundaries. Internal processes within the system do obey the Second Law: free energy, once expended, is unavailable to perform further work. But energy available to perform further work can be "imported" by open systems from their environment. There can be a transport of free energy—or negative entropy—across the system's boundaries.[2] When the two quantities—the free energy within the system, and the free energy transported across the system's boundaries from the environment—balance and offset each other, the system is in a steady (i.e., stationary) state. But in a dynamic environment the two terms seldom precisely balance over any extended period of time; therefore, the systems that evolve in the real world tend to fluctuate around certain states that define their dynamic steady states, rather than settle into them without variation.

The matter-energy systems that emerge in the course of evolutionary processes in the real world are systems in the third state, far from thermal and chemical equilibrium. Such systems evolve in the universe as in the course of time matter decouples from radiation and, in its uneven distribution in galaxies and stars, creates intense and enduring energy flows. (For details see the processes of cosmic evolution described in Chapter 3.)

The Continuum of Evolution

The processes of evolution unfold in all domains of the empirical world: in the physical universe the same as in the world of the living. They produce systems in the third state—dynamic matter energy systems far from equilibrium. These systems form a continuum that bridges the traditional boundaries of classical disciplines. This becomes evident if we compare systems that have emerged in the course of time in the universe and on earth in regard to some basic parameters. The relevant parameters include size, organizational level, bonding energy, and level of complexity.

Concentrating first on size, organizational level, and bonding energy, we find a truly elegant continuum. As we move from microscopic systems on a basic level of organization to macroscopic systems on higher organizational levels, we move from systems that are strongly and rigidly bonded to those with weaker and more flexible binding energies. Relatively small units with strong binding forces act as building blocks in the formation of larger and less strongly bound systems on higher levels of organization. These in turn become building blocks in still larger, higher-level, and less strongly bonded units.

The continuum of size, organizational level, and bonding energy appears throughout the range of evolution, from the basic particles of the universe to the highest-level systems of the biosphere. Particles such as the quark—although they do not exist as individual units in nature—are known to be bound by enormously high forces. Protons and neutrons within the nucleus of atoms

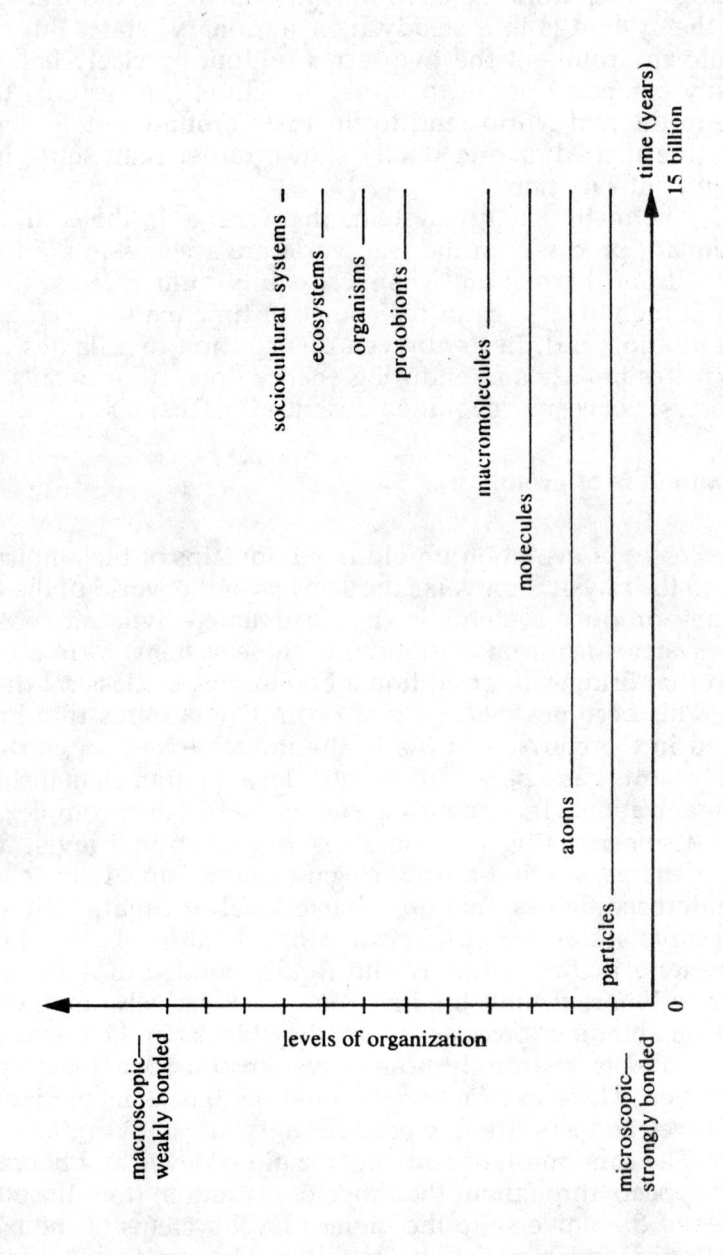

Figure 2.1. *The Size-Organization-Bonding Energy Continuum*

are bound by nuclear exchange forces, the strength of which is strikingly demonstrated in nuclear fission. The outer shell of atoms is bound to the nucleus by electronic bonding, an entire dimension weaker than the forces of the nucleus. Atoms within complex molecules are joined by ionic or covalent bonding and related weaker forces. The forces that join chemical molecules within organic macromolecules are weaker still, while those that bond cells within multicellular organisms are another dimension down the scale of bonding energy. Whatever the nature of the bonds that bind organic species and populations within ecologies and social systems, they are yet more ephemeral than physical and biochemical bonds.

Proportionately to the decrease of bonding energies we find an increase in level of organization. The concept of organization level is understood here in the sense of a Chinese-box structural hierarchy of "boxes within boxes." On a given level of organization, systems on the lower level function as subsystems; on the next higher level of organization, systems jointly form suprasystems. Using this concept, we can readily appreciate that the products of evolution are distributed on multiple hierarchical levels. Several particles jointly constitute atomic nuclei, and nuclei surrounded by electron shells form the atoms of the elements. Several atoms form simple chemical molecules, and more complex polymers are built from simpler molecules. Cells, in turn, are built from various kinds of macromolecules, organisms from cells, and ecologies and societies from populations and groups of individual organisms.

Evolution's level of organization does not determine the structural complexity of a system—the higher level is not necessarily more complex than its subsystems. For example, the structure of a molecule such as H_2O is considerably simpler than the atomic structure of hydrogen and oxygen. The structure of a cell colony is simpler than the structure of the constituent cells, and the structure of a termite colony, a baboon society, or an ecology is simpler than the organic structure of their individual members.[3]

There is no contradiction in the affirmation that the higher-level (supra)system is structurally simpler than the lower-level (sub)systems that are its parts. The hierarchy created by the evolution of systems is not only a structural hierarchy but, above all, a *control* hierarchy. In a control hierarchy the suprasystems control certain aspects of the behavior of the subsystems. Less complex systems on a higher level of organization can effectively control more complex systems on lower level in virtue of the selective disregard on the higher controlling level of the detailed dynamics of the lower-level units. As biophysicist Howard Pattee points out, the selective neglect of irrelevant details is a universal property of hierarchical control systems. Hierarchical controls always arise from internal constraints that force the lower-level units into a pattern of collective behavior that is independent of the details of their dynamic behavior. The emergence of a higher-level system is not a complexification, but a simplification of system function.

However, once a new hierarchical level has emerged, systems on the new level tend to become progressively more complex. For example, on the atomic level of organization, hydrogen, the first element to be synthesized in the processes of cosmic evolution, is structurally simpler than the subsequently synthesized heavier elements. On a higher level of organization, a molecule of water is simpler than a protein molecule; on a still higher organizational level, a unicellular organism is less complex than a multicellular one and, on the even higher, *social* level of organization, a colony of baboons is less complex than a human society. Thus, while a new level of organization means a simplification of system function, and of the corresponding system structure, it also means the initiation of a process of progressive structural and functional complexification.

Systems on a higher level of organization include the subsystems on subsidiary organizational levels. Thus the *total* structure of the higher-level suprasystem is more complex than any of its subsystems. This does not contradict the affirmation that the structure of the suprasystem, qua system on its own, defining organizational level, is simpler than the structure of its individual subsystems. A colony of baboons, for example, is structurally simpler by far than the genotype and the phenotype of the individual baboons that are its members. Yet the colony taken as an inclusive total system consists of all the individual baboons and all the cells, molecules, and atoms in their bodies, *plus* all the relations that exist between the baboons. Thus, while the colony per se is structurally and functionally simpler than individual baboons, taken as an inclusive total system it is more complex. (Henceforth, when speaking of the structure and function of a system, I shall mean by it the structure and the function of the system on its own organizational level, rather than the total system that includes the structures and the functions of all its subsystems.)

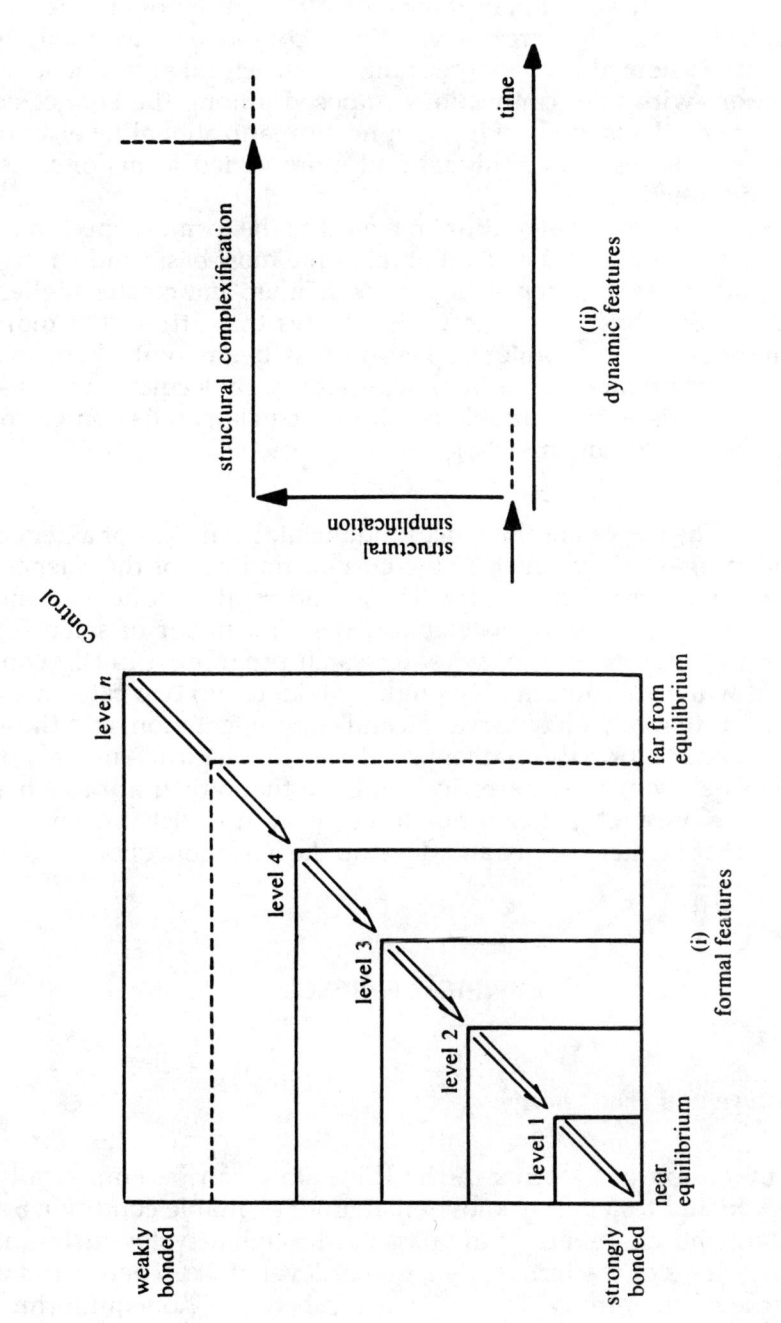

Figure 2.2. *Main Features of the Evolutionary Continuum*

A higher level of organization offers fresh possibilities for complexification; the greater variety of components available to the suprasystem allows a larger range of structural and functional variation, with new connections imposed among the connected subsystems. Thus, by moving to a new organizational level, evolution penetrates to ever higher and more varied forms of structure and function.

The logic of evolution in regard to the formal aspect of its products is simple and elegant: it takes the most basic and strongly bound systems, exposes them to each other, and creates higher-level systems based on the weaker forces that attract the more strongly bound components. The process begins with hadrons, leptons, and quarks; continues with atoms, molecules and cells; and it extends under suitable conditions (such as those on earth) all the way to organisms, ecologies, and societies.

The above are the most fundamental concepts of a general theory that brings together the current findings of the sciences of evolution and complexity. These and related concepts and hypotheses are tested and elaborated in a number of scientific fields and in a variety of ways. Research programs directly concerned with evolution can be roughly divided into two categories: programs that rely on observation and experimentation, and those that construct formal—mathematical and topological—models of system behavior. The empirical and the theoretical approaches complement each other; both have come up with important results that further specify and develop the basic concepts.

EMPIRICAL FINDINGS

Structure and Free Energy

The evolution of systems in the third state can be empirically observed and tested. It is known that, under suitable conditions, a constant and rich energy flow passing through a system drives it toward states characterized by a higher level of free energy and a lower level of entropy. The relationship between nonequilibrium and self-organization was pointed out by Prigogine as early as 1945. Then, experiments reported by Morowitz in 1968 demonstrated that a flow of energy passing through a nonequilibrium system organizes its structures and components to access, use, and store more and more free energy.[4] The explanation of this phenomenon in thermodynamic terms was given by Aharon Katchalsky in 1971: he showed that increasing energy penetration drives systems consisting of a large number of diffusely coupled nonlinear elements into nonequilibrium.

But the important measure in evolution is not simply energy, nor even free-energy, but *free energy flux density*. What counts is how much of the free-energy flux available in the environment is captured, retained, and used in a system. Free-energy flux density is a measure of the free energy per unit of time per unit of mass: for example, erg/second/gram. A complex chemical system retains more of this factor than a monatomic gas; a living system retains more than a complex chemical system. This indicates a basic direction in evolution, an overarching sweep that, together with the decrease of entropy and of equilibrium, defines the arrow of time in the physical as well as in the biological and the social world.

The relationship between energy flow over time and change in entropy and free energy is essential for answering not only the question as to *how* systems in the third state evolve, but also whether they evolve *necessarily*, under certain conditions. Until the 1970s investigators leaned to the view—exposed most eloquently by Jacques Monod—that evolution is due mainly to accidental factors. But as of the 1980s many scientists have become convinced that evolution is not an accident, but occurs necessarily whenever certain parametric conditions are fulfilled.

Laboratory experiments and quantitative formulations now corroborate the nonaccidental character of the processes of evolution. The experiments call for reproducing an energy flow from a source to a sink, and placing the test objects within the flow. As Prigogine points out, the best known experiment of this kind is that which demonstrates the appearance of the so called Bénard cells.

The experiment is basically simple. A vertical temperature gradient is created in a liquid by heating its lower surface. Thereby a continuous heat flux moves from the bottom toward the top. When the gradient reaches a critical value, the state of the liquid in which heat is conveyed upward by conduction becomes unstable. A convection flow is created, increasing the rate of heat transfer. The flow is in the form of a complex spatial organization of molecules in the liquid: moving coherently, they form hexagonal cells of a specific size. The Bénard cells maintain themselves in the heat flow and they dissipate entropy at a high rate. The phenomenon is radically different from the linear processes that occur in systems near equilibrium.

Bénard cells are not limited to laboratory experiments; they occur in nature in a wide variety of circumstances. Wherever there is a complex flow in a medium, forms of new order arise spontaneously. This is true of the flow of heat from the center of a star such as the sun to its outer layers, as well as the flow of warm air from the surface of the earth toward outer space. The earth, warmed by the sun, heats the air from below while outer space, much colder, absorbs heat from the top layers of the atmosphere. As the lower layer of the atmosphere rises and the upper layer drops, circulation vortices are created, and these take the shape of Bénard cells. Closely packed hexagonal lattices, such cells leave their imprint on the pattern of sand dunes in the desert and on the pattern of snowfields in the Arctic. Surrounding us in the many complex flows of the biosphere, Bénard cells are evidence of systemic self-organization in nature: they represent the kind of order encountered in all third-state systems far from thermodynamic equilibrium, whether living or nonliving.

Basic experiments in inorganic chemistry shed light on how systems of interacting elements move from states at or near equilibrium to the domains of nonequilibrium that characterize systems in the third state. Under controlled laboratory conditions sets of chemical reactions can be forced to move progressively further from chemical equilibrium. Near chemical equilibrium the reaction system is successfully described by solving the chemical kinetic equations that apply at equilibrium as well as those that correspond to the Brownian motion of the molecules and the random mixing of the components. But as reaction rates are increased, at some point the system becomes unstable and new solutions are required to explain its state, branching off from those that apply near equilibrium. The modified solutions signify new states of organization in the system of reactants: stationary or dynamic patterns of structure, or chemical clocks of various frequency. Where the equilibrium branch of the solution becomes unstable the reaction system acquires characteristics typical of nonequilibrium systems in general: coherent behavior appears, and a higher level of autonomy vis-à-vis the environment. The elements cohere into an identifiable unity with a characteristic spatial and temporal order: there is now a dynamic *system*, whereas near equilibrium there were but sets of reactants.

The origins of life on earth may be due to just such a process of progressive organization in systems of continuous energy flow. It has been known for some time that steady irradiation by energy from the sun was instrumental in catalyzing the basic reactions which led to the first protobionts in shallow primeval seas. In this process the energy of the earth itself may have played an essential role, in the form of hot submarine springs in the Archaean oceans. No more than a series of small continuous flow reactors may have been required, similar in all essential respects to the chemical reaction systems responsible for Bénard cells, as well as for more complex chemical structures under controlled laboratory conditions. Nature's reaction systems may have consisted of cracking fronts in the submarine rock which caused sea water to heat rapidly and to react with chemicals in the rock and in the surrounding sea water. As the hot fluid rose toward the surface it dissipated its heat. In the continuous flux of energy created by magma erupting into the sea bottom and interacting with the energy of the sun in shallow seas and then dissipating into air and water, the chemicals that served as the basic building blocks of life were constantly mixed. Out of these series of reactions came the protobionts—the lipid vesicles from which more complex forms of life have evolved in time.

Systems with significant unity, autonomy, and ordered structure and behavior emerge when sets of reactants are exposed to an energy flow. If there is sufficient diversity in the components and sufficient complexity in their structure, the resulting system will have bistability or multistability—that is, it will be capable of persisting in more than one steady state. In more complex systems there will also be various feedbacks and catalytic cycles among the subsystems or principal components.

In nature, systems in the third state almost always exhibit some variety of catalytic cycles. Such cycles tend to be selected in the course of time in virtue of their remarkable capacity for persistence under a wide range of conditions. They have great stability and fast reaction rates. Already in 1931 Lars Onsager could show that in a steady-state system, cyclic matter-energy flows are likely to arise. For example, in a simple chemical system composed of three types of molecules, A, B, and C, in which both forward and reverse reactions are possible (e.g., $A \rightleftharpoons B$, $B \rightleftharpoons C$, $C \rightleftharpoons A$) the introduction of a continuous energy irradiation into one of the cycles, (e.g., $A + hv \rightarrow B$) tends to move the system into a cyclic pattern ($A \rightarrow B \rightarrow C \rightarrow A$). More recently Manfred Eigen and Peter Schuster have demonstrated that catalytic cycles are the basis for the persistence of the complex structures which underlie, and make possible, the emergence of life.

In relatively simple chemical systems, autocatalytic reactions tend to dominate, while in more complex processes, characteristic of living phenomena, entire chains of cross-catalytic cycles appear.[5] Catalytic cycles underlie the stability of the sequence of nucleic acids that code the structure of living organisms; they also underlie, on a higher organizational level, the persistence of all living forms in the biosphere of our planet.

Convergence to Higher Levels

Given sufficient time, and an enduring energy flow acting on organized systems within permissible parameters of intensity, temperature, and concentration, the already evolved catalytic cycles tend to interlock in what Eigen calls hypercycles. These are cycles that maintain two or more dynamic systems in a shared environment through coordinated functions, similar to but even more integrated than symbiosis among organic species. For example, on the basic level of living phenomena, nucleic acid molecules carry the information needed to reproduce themselves as well as an enzyme. The enzyme catalyzes the production of another nucleic acid molecule, which in turn reproduces itself plus another enzyme. The loop may involve a large number of elements; ultimately it closes in on itself, forming a cross-catalytic hypercycle remarkable for its fast reaction rates and stability under diverse parametric conditions.

The formation of hypercycles allows dynamic systems to emerge on successively higher levels of organization. On the new level the amount of information that can be handled by the cycle is greater than on lower levels owing to the greater diversity and richness of the components and structures. Evolution thus bypasses the functional optima of complexity that would block further progress on given organizational levels. Instead of drifting randomly around the optimal complexity on level n, the formation of hypercycles allows two or more systems to jointly create a suprasystem on level $n + 1$, where fresh structural possibilities offer new horizons for further evolution.

The shift from level to level of organization by means of catalytic hypercycles we can best call *convergence*. (*Hypercycle formation* or *hierarchization* would be correct as well, though these expressions are less clear and clumsier in use.) In this context convergence does not mean growing similarities among systems (as

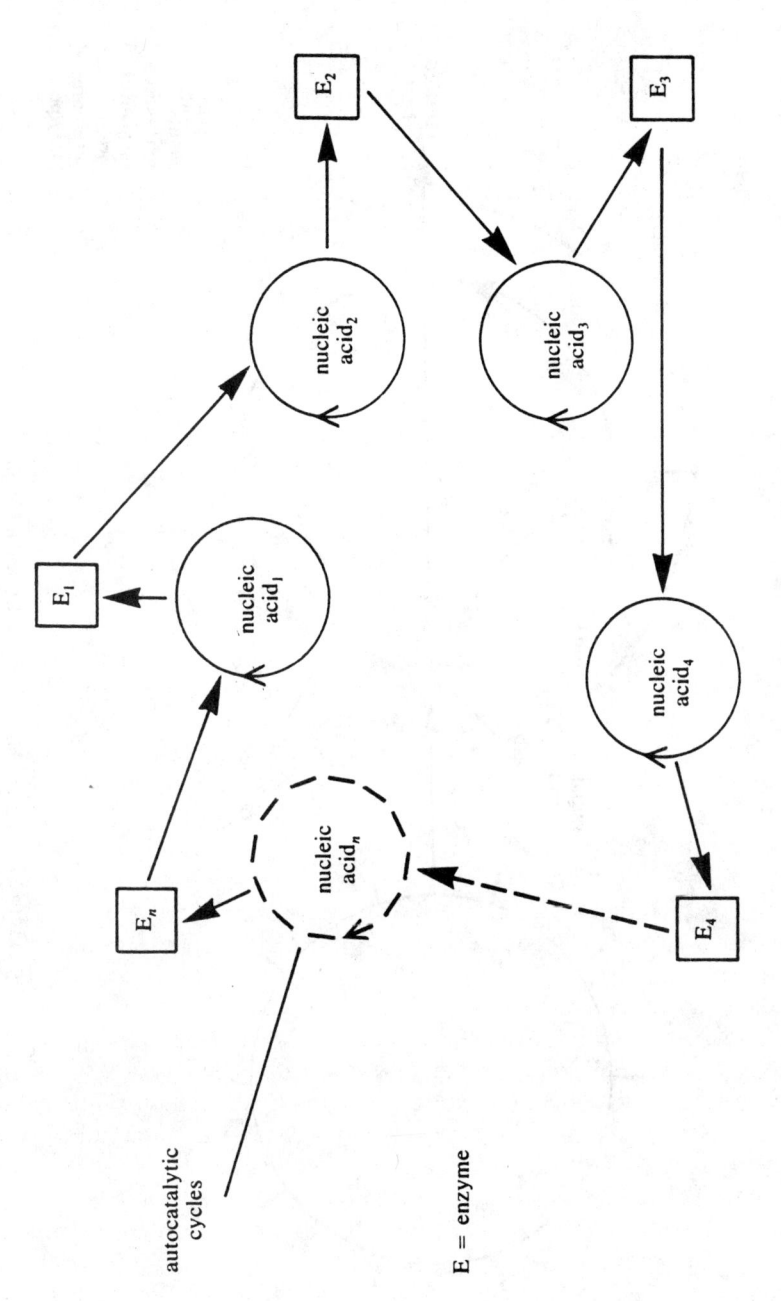

Figure 2.3. A Cross-Catalytic ("Hyper") Circle

36   THE LOGIC OF EVOLUTION

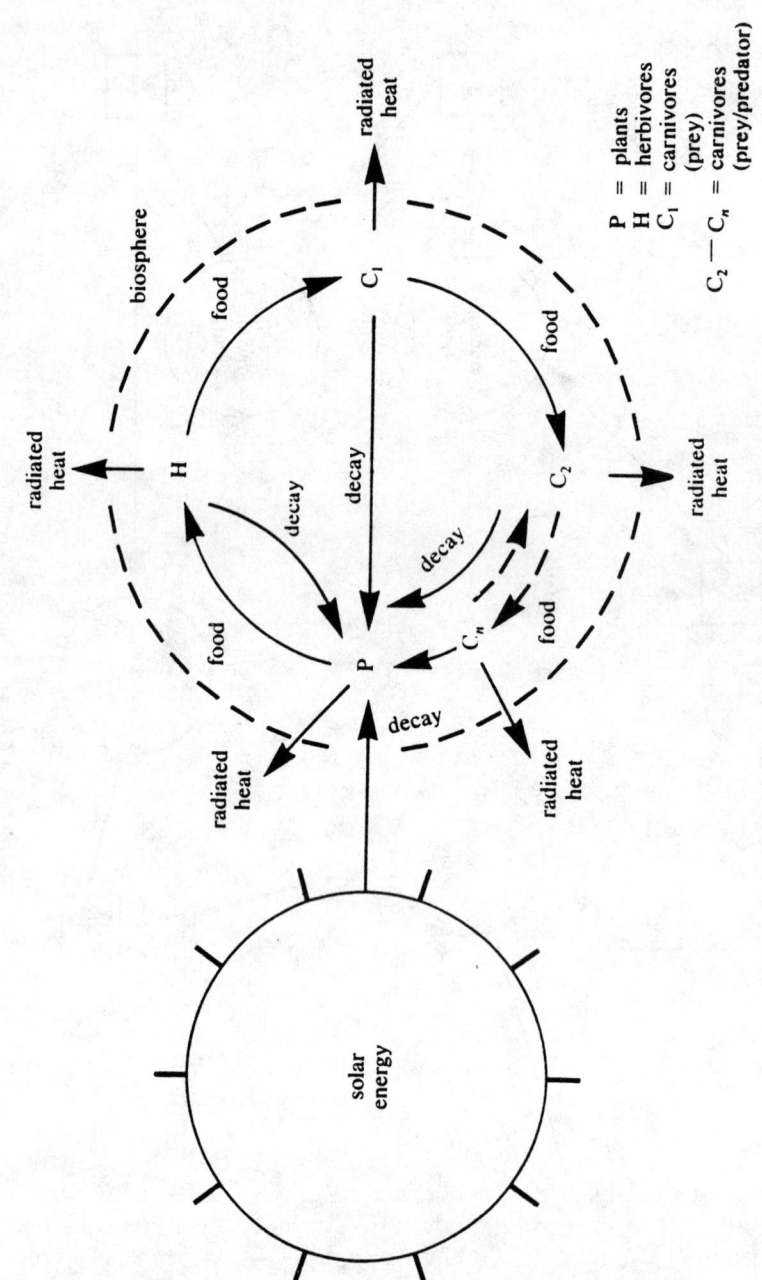

Figure 2.4. Some Major Catalytic Cycles in the Biosphere

in the convergence of ideologies and socioeconomic systems) since the hypercycle-forming systems functionally complete and complement each other. The outcome of convergence is a higher-level system that selectively disregards many details of the dynamics of its subsystems and imposes an internal constraint that forces the systems into a collective mode of functioning.

Convergence among systems occurs in all realms of evolution. On each level, systems in an energy flow progressively exploit the free-energy fluxes in their environment. As the density of the free energy retained in the systems increases, the systems acquire structural complexity. Because of their convergence, systems on successively higher levels of organization acquire a fresh range of possibilities for setting forth the process by which they retain increasingly dense and abundant free-energy fluxes in structures of growing complexity.

The processes of evolution create systems on multiple hierarchical levels. On each level the structure of the highest system level is initially comparatively simple: simpler than the structure of the component subsystems. (Of course, the complexity of the suprasystem-cum-subsystems is more complex than any of its subsystems: it includes all subsystemic structures plus the relations between them.) The further evolution of the suprasystem then leads to the progressive complexification of its defining system level and ultimately to the creation of hypercycles that shift it to the next organizational level. Thus evolution moves from the simpler to the more complex type of system, and from the lower to the higher level of organization.[6]

Instability and Transformation

Change occurs and evolution unfolds because dynamic systems in the third state are not entirely stable. They have upper thresholds of stability that, if transgressed, produce critical instabilities. Experiments confirm that dynamic systems far from equilibrium can be moved out of their steady states by changes in some of the environmental parameters. Such systems are critically sensitive to changes in the parameters that are essential for the functioning of the structure-maintaining catalytic cycles. When changes of this kind occur, the systems enter a transitory phase characterized by indeterminacy, randomness, and some degree of chaos, with a relatively high rate of entropy production. The chaotic phase comes to an end when the systems settle into a new dynamic regime. In this new ordered state the systems are again maintained by catalytic cycles and multiple feedbacks, and entropy production drops to a functional minimum.

The way in which systems in the third state respond to destabilizing changes in their environment is of great importance for understanding the dynamics of evolution in all realms of reality. As Part Two will show, there is significant evidence accumulated in many fields of empirical science that dynamic systems do not evolve smoothly and continuously over time, but do so in comparatively sudden leaps and bursts. Third-state systems can evolve through sequences of destabilizations and phases of chaos since they have multiple states of stability: when one steady state is fatally disturbed, other steady-state solutions remain accessible. The further systems are from thermodynamic equilibrium, the more sensitive to change their structure and the more sophisticated the feedbacks and catalytic cycles that maintain it. Complex systems far from equilibrium can exist in many different steady states. The number of possible steady state solutions grows proportionately to their complexity and level of nonequilibrium.

The selection from among the set of dynamically functional alternative steady states is not predetermined. The new state is decided neither by initial conditions in the system, nor by changes in the critical values of environmental parameters; when a dynamic system is fundamentally destabilized it acts indeterminately. By an apparently random choice, one among its possibly numerous internal fluctuations is amplified, then the fluctuation spreads with great rapidity. The thus "nucleated" fluctuation dominates the system's dynamic regime and determines the nature of its new steady state. Since the number of the available steady states increases proportionately to the level of nonequilibrium, the more dynamic and negentropic the system the greater its degree of freedom in the wake of its destabilization.

Owing to the factor of chance that thus enters the processes of transformation, the evolutionary trajectory of systems in the third state is not predictable in detail. Two systems, even if they start from identical initial conditions and are subject to identical perturbations in the environment, may evolve along divergent trajectories. We cannot predict the precise evolutionary trajectory chosen by any system in the third state; all we can say is that the new steady state, if it is dynamically stable, assimilates the disturbance introduced by the destabilizing parameters within an open structure that is likely to be more dynamic and complex than the structure in the previous steady state.

OUTLINE OF THE GENERAL THEORY

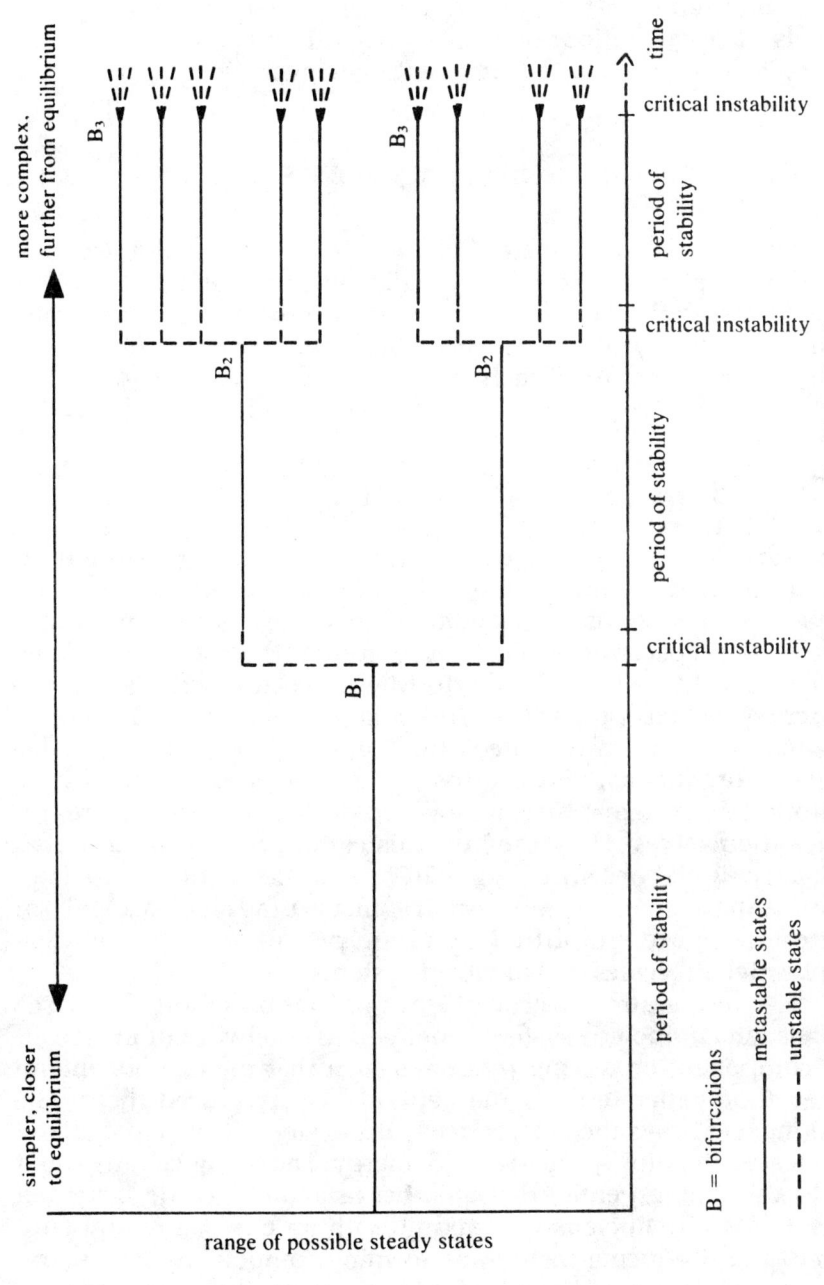

Figure 2.5. *Bifurcations: The Increase of Alternatives at Higher Levels*

It is through phases of chance and indeterminacy that evolution tends in the observed general direction of successive levels of organization, with growing dynamism and increasing complexity on each of the organizational levels.

THEORETICAL FINDINGS

A detailed understanding of the observed dynamics of the evolution of systems in the third state calls for the development of new theoretical tools. This is true in regard to both the self-maintenance of such systems in their environment, and the discontinuous, nonlinear nature of their evolution and development.

Autopoiesis

Systems on organizational levels where life emerges are too far from thermodynamic equilibrium to persist indefinitely in their milieu. Such systems can maintain themselves in time only if they evolve the capacity to replicate or reproduce their structure. Using a term introduced by Humberto Maturana in the 1960s, we can describe systems on such relatively high organizational levels as "autopoietic" (from the Greek for "self-creating"). Indeed, cells, organs, organisms, and groups and societies of organisms are autopoietic systems: they renew, repair, and replicate or reproduce themselves. How they do this cannot be known in detail (already a cell consists of some 10^{20} molecules with hundreds of simultaneous reactions), but it can be modeled. Models of autopoiesis are simplified dynamic paraphrases of complex replicative processes in real-world systems.

In the already classic work of Maturana and Francisco Varela, an autopoietic system is defined as a network of interrelated component-producing processes such that the components in interaction generate the same network that produced them. The product is always the network (i.e., the system) itself, created and re-created in a flow of matter and energy. The living cell turns out to be the simplest entity that qualifies as an autopoietic system. It has a catalytic nucleus interacting with an environmental substrate and producing membrane-forming components. The membrane separates the network of interactions that produced it from the rest of the cell. Subcellular structures are defined in the autopoietic model as component-producing components that replicate the network of self-replicating processes.[7] The elements of the model can be computerized, and investigators can simulate the development of various types of self-replicating systems. Since the substrate inputs can be controlled and the rates of membrane formation and disintegration can be regulated, myriad variations can be induced. Systems with rupture-proof membranes can be produced with equal facility as systems that are incapable of forming membranes; one variety of systems proves to be self-repairing while another turns out to have a cancerous growth. Investigators hope that some of the simulations may prove to be faithful paraphrases of self-replicating processes not only in cells, but also in organs, in multicellular organisms, and in ecological and social systems.

Autopoietic modeling evolved from the rigorous mathematical base of cellular automata theory pioneered by John von Neumann in the 1960s. Today autopoietic models and cellular automata theory are in rapid evolution, thanks to ever more powerful computers. A great variety of self-replicating systems now pursue their fascinating and often unpredictable developmental cycles in the electronic workshops of mathematicians. It is often difficult to say which of these systems simulate real-world systems; a question rendered even more opaque by the lack of interest many theoreticians profess in the practical application of their models.

Autopoiesis is not evolution, however, even if autopoietic cellular automata models can simulate certain evolutionary phenomena such as the convergence of cells into multicellular systems. In order to understand evolution one needs also to understand how *discontinuous, nonlinear* change takes place in real-world dynamic systems. This calls for different types of models, and even for a new mathematics. (The differential calculus, ordinarily employed to model change, assumes that change is relatively smooth and continuous; it becomes almost unmanageably complex when forced to handle leaps and discontinuities.) It has been dynamic systems theory that has risen to meet this challenge. The models of dynamic systems theorists can likewise be computerized, but they consist not of components with specially designed functions, but of a combination of ordinary differential equations, partial differential equations of the evolution type, and finite difference equations singly or in sets.

Catastrophes, Chaos, and Bifurcations

Although the above nomenclature of dynamic systems theorists is decidedly pessimistic, their work is hopeful: it promises to shed new light on the nonlinear trajectories that characterize the evolution of complex systems in the real world. René Thom, a pioneer of dynamic systems theory, has developed multidimensional topological models that permit a rigorous treatment of the succession of system states even if the pattern of change is fundamental and sudden, that is, if it is "catastrophic"—hence the name "catastrophe theory."[8]

Catastrophe theory provides a mathematical tool for the rigorous representation of certain varieties of discontinuous change. However, Thom's standard theory can only model one of a larger variety of radical system changes, namely changes due to the sudden appearance or disappearances of so-called static attractors. Recent work in dynamic systems theory has led to the development of quantitative tools to model catastrophic changes due to the appearance or disappearance of other types of attractors as well, including those known as "periodic" and "chaotic."

In the geometric orientation that dominates in contemporary dynamic systems theory, the principal features of dynamic systems are the attractors: they characterize the long-run behavior of the systems. Dynamic systems evolve from a given initial state along a unique trajectory (or time series) of states in accordance with the laws of evolution. This leads eventually to some recognizable pattern where the trajectory remains trapped. The pattern defines the *attractors* of the system, while the ensemble of the initial states that are attracted by them define their *basin*.

If the series of system states comes to a rest, their evolution is governed by a static attractor; if the states consist of a repeated cycle of states with a definite periodicity, the system is under the sway of a periodic attractor. And if the trajectory of system states neither comes to rest nor exhibits periodicity but is highly erratic, it is under the influence of a so-called chaotic attractor.

In recent years chaotic behavior has been discovered in a wide variety of natural systems, and their mathematical modeling has made rapid progress. An entire discipline has sprung up within dynamic systems theory devoted to the study of the properties of chaotic attractors and of the systems governed by them; it became popularly known as chaos theory. Despite its name, the theory seeks to eliminate rather than discover or create chaos—it

OUTLINE OF THE GENERAL THEORY 43

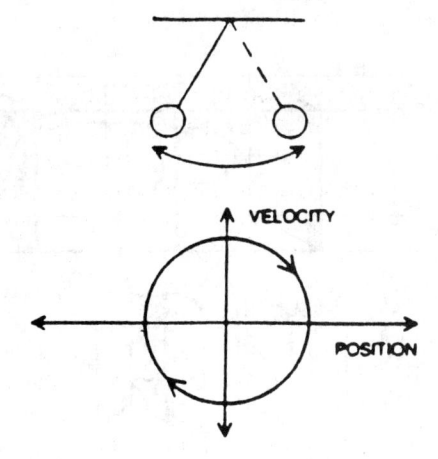

a. Motion can be represented graphically in the so-called phase space. Above, the movement of an imaginary frictionless pendulum, starting from a given set of initial conditions. The diagram shows how the velocity and the position of the bob on the pendulum change together, indicating all the coupled velocity-positions generated by the system.

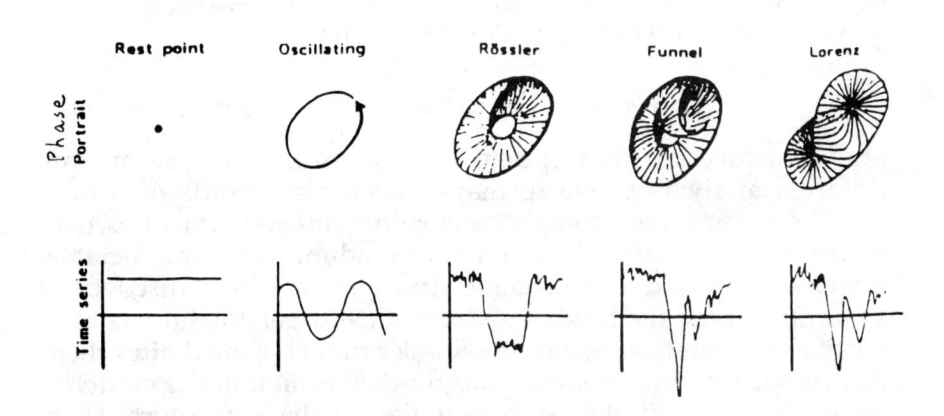

b. More complex behavior can also be represented in the phase-space, using the device of attractors. Such representation can exhibit order, where the usual time series diagram (showing how a given parameter varies over time) is seemingly disordered, as in the case of representation by the so-called chaotic attractors (here illustrated by the Rössler band, the Funnel, and the Lorenz attractor).

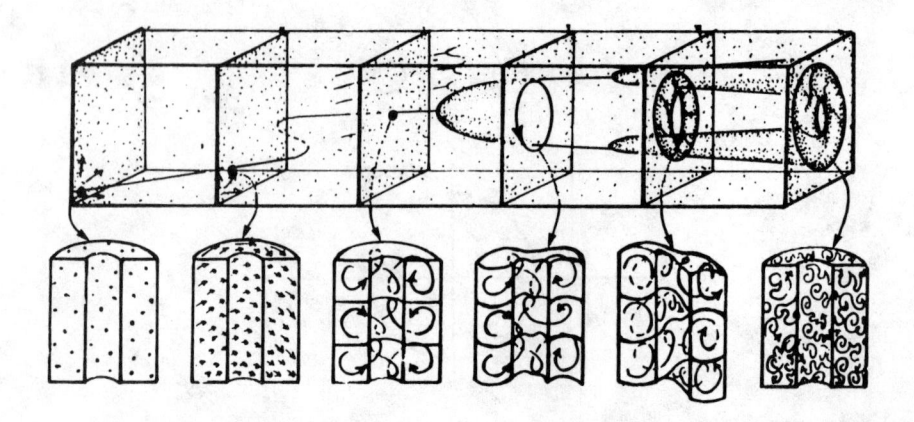

Representations in phase-portraits model real-world systems and exhibit order where a "snapshot"-type representation would only show chaos. Above: the movement of liquid in a spherical container. Each cut-away snapshot of the container corresponds to one phase-portrait showing the evolution of the attractors. The point-attractor shows imaginary ink-drops in the liquid at rest; as the point attractor evolves, more complex movement is registered in the container, until the movement becomes entirely chaotic (on the right). Nevertheless, the chaotic attractor (a 'donut') exhibits the corresponding form of order.

Figure 2.6 (a-c). *The graphic representation of chaos*

studies processes that appear chaotic on the surface but on detailed analysis prove to manifest subtle strands of order. Chaotic attractors are complex and subtly ordered structures that constrain the behavior of seemingly random and unpredictable systems. Several kinds of chaotic attractors have been discovered so far (including the Birkhof and the Shaw bagel, the Lorenz butterfly, the Rössler band; and the Rössler funnel, named after their discoverers and the geometric shapes they exhibit in the models); many more are expected to come to light in the near future. They reduce seeming chaos to complex varieties of order in processes as varied as fluids in flow and the blending of substances during solidification. The phenomenon of turbulence is a case in point: it has been known since the nineteenth century, but its origins have been imperfectly understood. By 1923 experiments in fluid dynamics had demonstrated the appearance of annular Taylor vortices; these appear when the speed of stirring in a fluid increases beyond a critical point. Further increases in stirring produce additional abrupt transformations and ultimately turbulence. Turbulence, however, can be modeled by chaos theory and proves not to be entirely chaotic: it exhibits the complex varieties of order inherent in systems governed by chaotic attractors.

The behavior of dynamic systems in the empirical world is normally influenced by many different attractors simultaneously; dynamic systems theory seeks to account for such systems by correspondingly complex models that, for the most part, are yet to be developed. But even the existing models simulate the behavior of real-world systems with surprising fidelity. The key feature of the behavior of the models is the analysis of the sudden phase change that leads empirical systems from one kind of stable state to another. In dynamic systems models, such changes in systems behavior represent bifurcations of the trajectory traced by the system states. Bifurcations appear in the "phase portrait" of the systems and consist of a shift from one type of attractor to another. In a shift from a static to a periodic attractor, a hitherto stable system begins to oscillate, while in a shift from periodic to chaotic attractors, a previously oscillating system lapses into chaos. These so-called subtle bifurcations are but one variety of basic system changes; the other variety is known as catastrophic bifurcations and consists of the sudden appearance or disappearance ("out of the blue") of static, periodic or chaotic attractors.

Bifurcations in the complex systems mapped by dynamic systems theory have direct application to real-world systems. Subtle bifurcations can model increasing instability in third-state systems far from thermodynamic equilibrium. A system, such as a series of chemical reactions in stable equilibrium, begins to oscillate; or an oscillating system, such as a "chemical clock," becomes turbulent. In its mathematic models, dynamic systems theory identifies a number of such "scenarios" leading from stable equilibrium to chaos.

Models with catastrophic bifurcations (conducing from turbulent to newly ordered states through the reconfiguration of the attractors) simulate rapid evolutionary leaps with the greatest fidelity. The significant simulations occur when dynamic systems are destabilized and pass through a chaotic phase on the way toward essentially new—and in practice unpredictable—steady states. The chaotic phase is not entirely random even if it is highly erratic and unpredictable: it is governed by chaotic attractors. During this phase the bifurcating systems are sensitive to minute changes: the smallest variation in an initial condition can give rise to widely differing outcomes (Dynamic systems theorists speak of "initial condition dependence" and illustrate this condition by interpreting the Lorenz butterfly—a pioneer chaotic attractor discovered by meteorologist Edwart Lorenz in 1963—as a metaphorical "butterfly effect": if a monarch butterfly flaps its wings in Southern California, a month later Outer Mongolia's weather becomes unpredictably different. Since the world's weather is in a permanently chaotic state, the story is only slightly exaggerated.)

As the explorations of the empirical realms of evolution in Part Two will show, subtle or catastrophic bifurcations, involving transitory chaotic phases in shifts from stable to chaotic and back to stable attractors, are the kind of transformations that underlie the evolution of all third-state systems in the real world, from atoms of the elements to societies of human beings.

SUMMARY

The main features of GET—general evolution theory—can be briefly summarized. Autocatalytic and cross-catalytic feedback loops predominate in open dynamic systems in the third state far from equilibrium in virtue of their rapid reaction rates and great stability. However, as no autopoietic reaction cycle is entirely immune to disruption, constant changes in the environment sooner or later produce conditions under which certain cycles can no longer operate. The systems encounter a point known in dynamic systems theory as a bifurcation. The outcome in complex systems, as both experiment and theory demonstrate, is essentially indeterminate: it is not a function of initial conditions. Bifurcations produce a selection from among the set of alternative steady states available to the system through the amplification of some existing fluctuations. The new steady states assimilate the perturbations within the limits of error-tolerance of the catalytic cycles and are likely to produce increases in the complexity of system structure. There is a significant probability that the system leaps to a new plateau of nonequilibrium—that it retains a greater quantity of a more dense flux of free energy for a longer time and thus becomes more dynamic and more autonomous in its milieu.

New levels of organization are attained in the system-forming process of evolutionary convergence as catalytic cycles on one level interlock and form hypercycles on the next. Thus molecules emerge from combinations of chemically active atoms; protocells emerge from sequences of complex molecules; eukaryotic cells emerge among the prokaryotes; metazoa make their

OUTLINE OF THE GENERAL THEORY 47

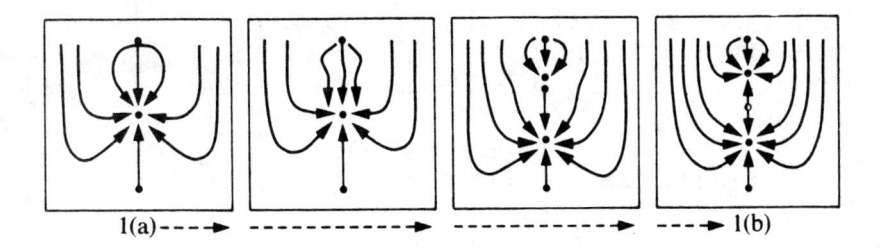

Figure 2.7(a)

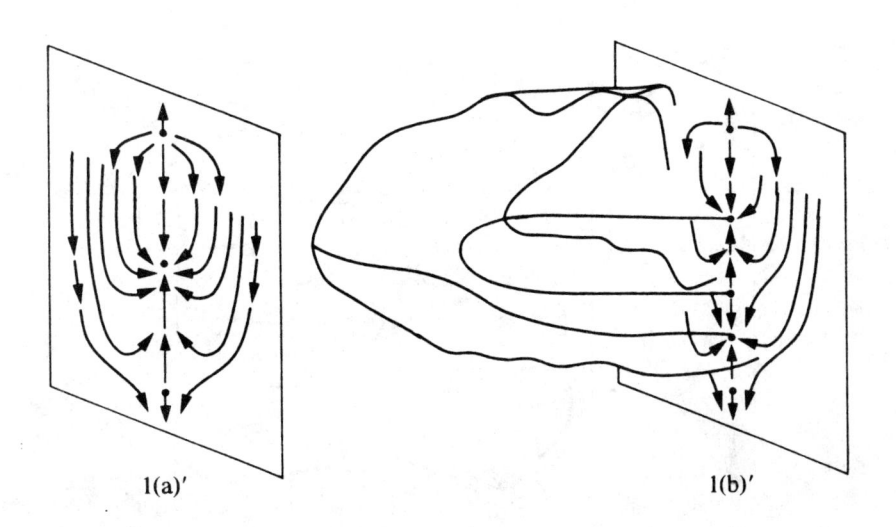

Figure 2.7(b)

In Figure 2.7 (a-b) the same catastrophic bifurcation is represented in a two- and a three-dimensional state space. The state space is a geometrical model for the virtual states of the system. Each point in the state space represents a single, instantaneous state. As each state is characterized by a unique evolutionary tendency described by the velocity vector, a given initial state evolves along a unique trajectory. Following a transient response phase, the trajectory asymptotically approaches a limit set which is the system's attractor. The attractor may be static, periodic, or chaotic. As 2.7(a) shows, the second static attractor appears "out of the blue," far from any other attractor (separatrices, the divisions between the basins of the attractors, are ignored for the sake of clarity).

2(a)   - - - ▶   2(b)

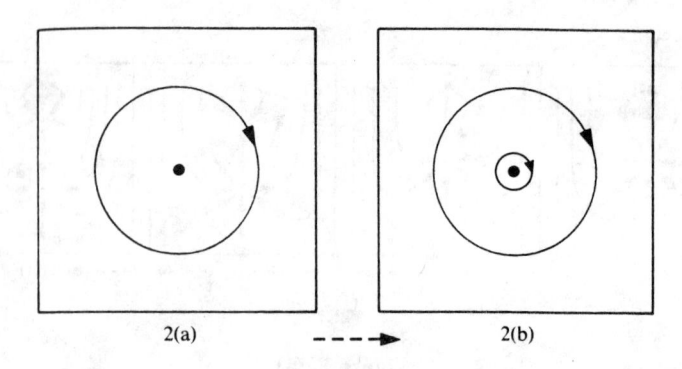

Figure 2.8(a)

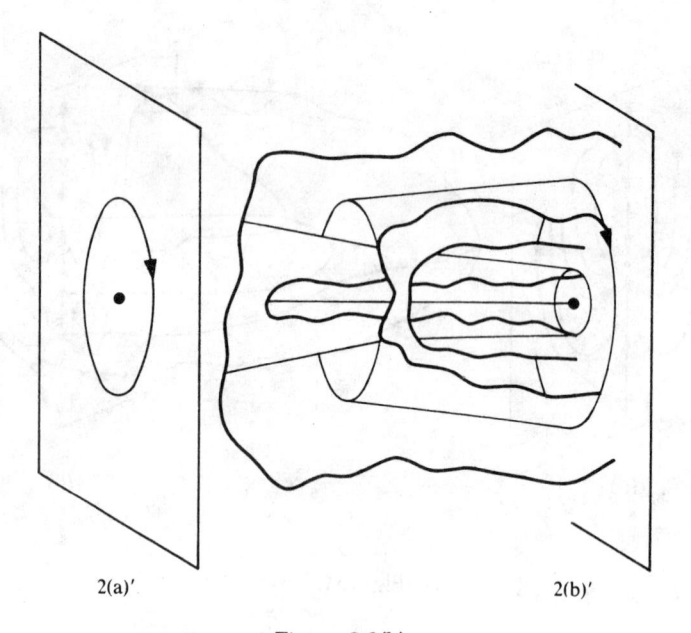

2(a)'   2(b)'

Figure 2.8(b)

Figure 2.8 shows a catastrophic bifurcation due to the sudden appearance of a periodic attractor. First one periodic attractor dominates the dynamics of the system in the center of its basin [2(a)]. Then, as the control parameters are changed, another periodic attractor appears out of the blue [2(b)]. Figure 2.8 (a-b) shows the same process in a three-dimensional state space, also without separatrices. (For details, see Abraham and Shaw, "Dynamics—The Geometry of Behavior." 1988, Aerial Press, Inc., Santa Cruz, CA.

OUTLINE OF THE GENERAL THEORY

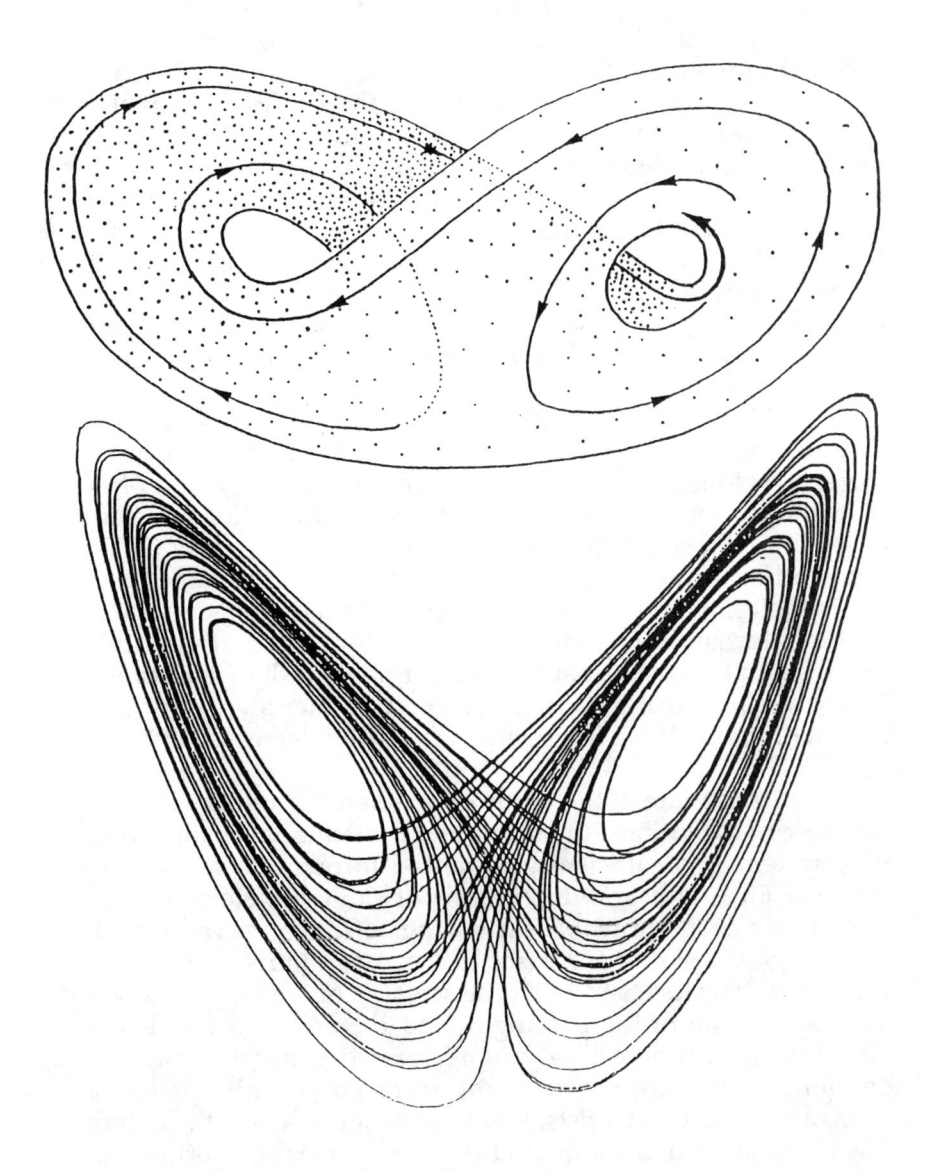

Figure 2.9. *The vaguely butterfly-shaped computer model of air currents in the atmosphere by meteorologist Lorenz, to which this attractor owes its name (above), and a more recent computer-drawn butterfly attractor (below). The trajectory of the system-states in the "phase portraits" is highly sensitive: the smallest perturbation can flip the system from one wing of the butterfly to the other.*

appearance among the protozoa, and metazoa join in more embracing ecological, social and, at least in one instance also sociocultural, systems.

Continuity and consistency in the processes of evolution give rise to continuity and consistency in its products. Periodic destabilizations of systems far from equilibrium, coupled with the organizing effect of a constant energy flow, push bifurcating systems up the ladder of the evolutionary hierarchy. This process adds weaker and more flexible bonding energies to stronger and more rigid ones, complexifies system structures on each organizational level and simplifies them again on the next level. It superposes highly developed control capabilities on more primitive forms of self-regulation.

The fact that systems far from equilibrium evolve through periodically indeterminate nonlinear transformations toward regions ever further removed from equilibrium offers a first glimpse of the foundations of general evolutionary theories based on valid scientific concepts. The main elements of GETs are nonequilibrium systems within an energy flow maintained by catalytic cycles, the mix of determinacy and randomness introduced by the alternation of ordered dynamic stability with periods of indeterminate transformation, and the statistically significant tendency toward greater complexity and a higher level of organization, with more dense free energy fluxes and decreased entropy.

In the late 20th century the scientific understanding of the logic of evolution—the forms and dynamics of change in complex systems—is sufficiently advanced to permit a revival of the classical ideal of a grand synthesis of the main elements of our knowledge of the principal furnishings of the observed world. In our day the evolutionary paradigm is ready to move from metaphysics to physics—and to all the sciences of the empirical world. The exploration of the paradigm, and the creation, criticism and elaboration of progressively more refined general theories, is a challenge awaiting the contemporary community of natural, human and social scientists. It is a challenge well worth accepting. The elaboration of a sound and detailed general evolution theory will surely rank among the greatest achievements of human intellectual history.

OUTLINE OF THE GENERAL THEORY

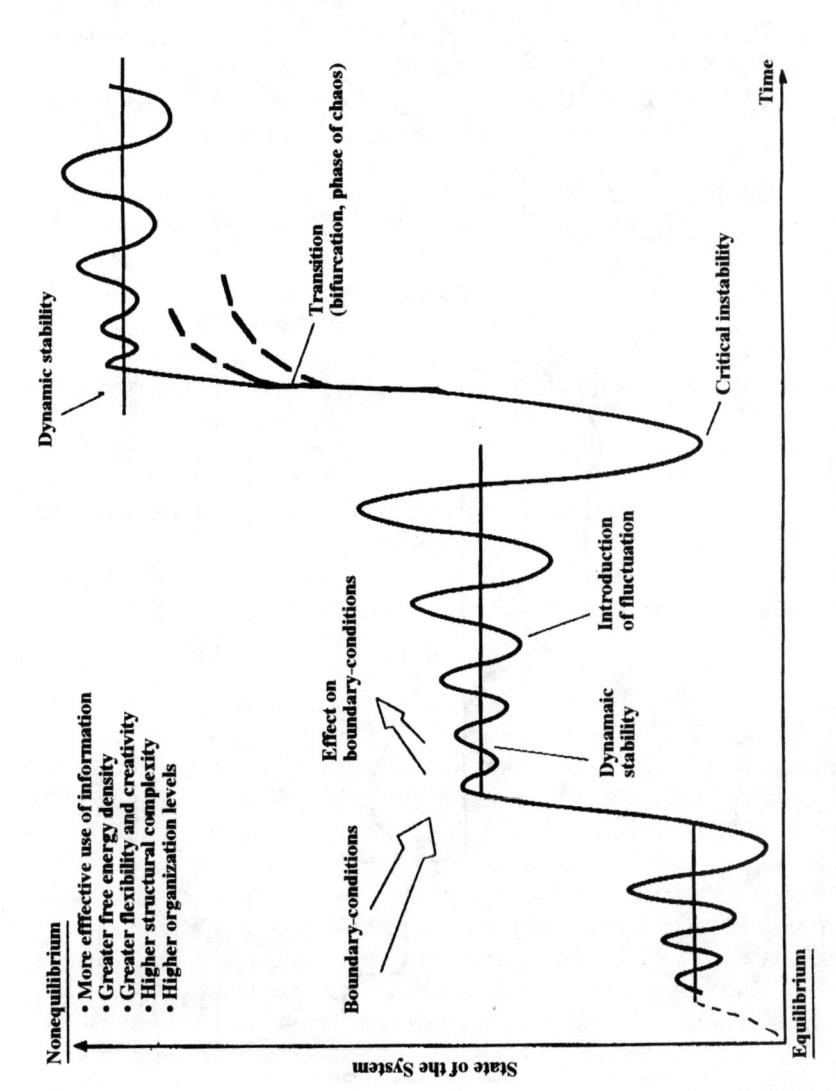

Figure 2.9. *The statistically one-way buildup of dynamic nonequilibrium systems*

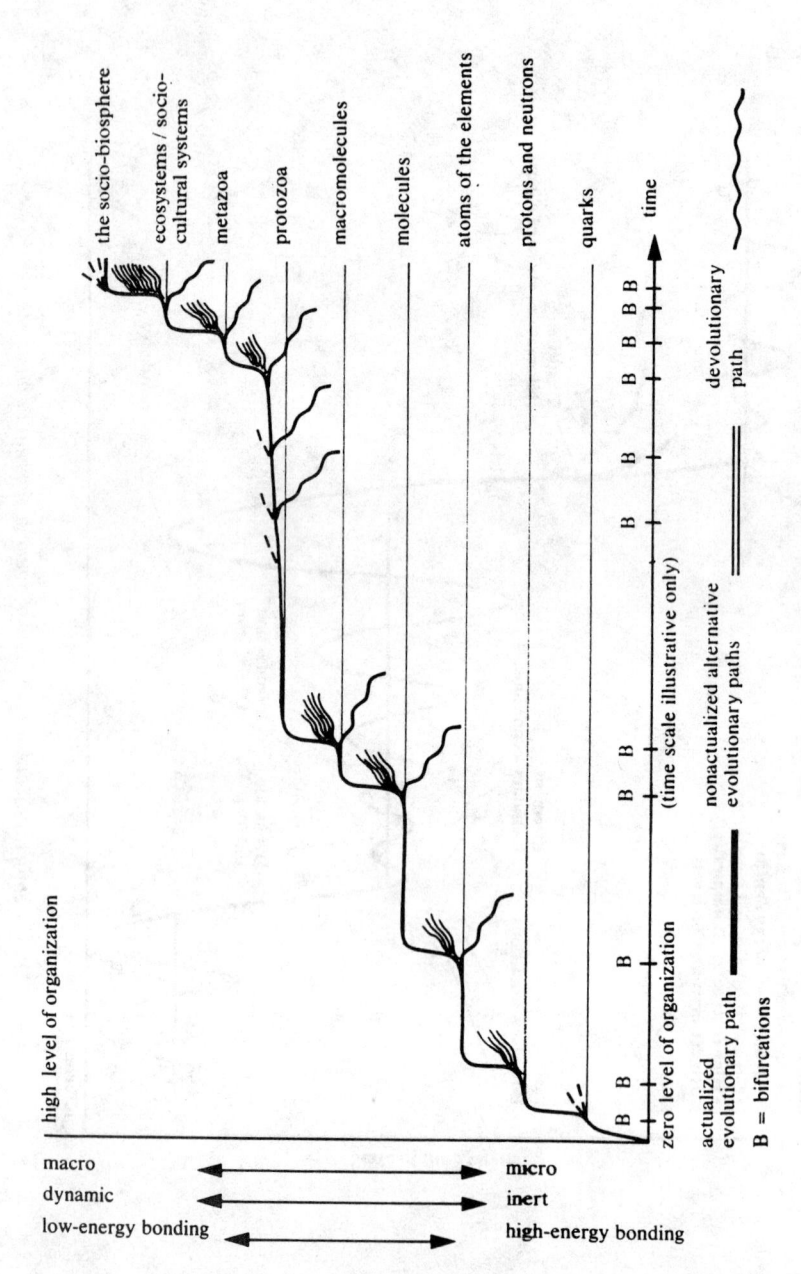

Figure 2.10. *The emergence of successive levels of organization through convergence*

Notes

1. In a condition of equilibrium the production of entropy, and forces and fluxes (the rates of irreversible processes), are all at zero, while in states near equilibrium entropy production is small, the forces are weak, and the fluxes are linear functions of the forces. Thus a state near equilibrium is one of *linear* nonequilibrium, described by linear thermodynamics in terms of statistically predictable behaviors, as the system tends toward the maximum dissipation of free energy and the highest level of entropy. Whatever the initial conditions, the system will ultimately reach a state characterized by the least free energy and the maximum of entropy compatible with its boundary conditions.

2. Change in the entropy of the systems is defined by the well-known Prigogine equation $dS = d_iS + d_eS$. Here dS is the total change of entropy in the system, while d_iS is the entropy change produced by irreversible processes within it and d_eS is the entropy transported across the system boundaries. In an isolated system dS is always positive, for it is uniquely determined by d_iS. which necessarily grows as the system performs work. However, in an open system d_eS can offset the entropy produced within the system and may even exceed it. Thus dS in an open system need not be positive: it can be zero or negative. The open system can be in a stationary state ($dS = 0$), or it can grow and complexify ($dS < 0$). Entropy change in such a system is given by the equation ($d_eS = -d_iS \leq 0$); that is, the entropy produced by irreversible processes within the system is shifted into the environment.

3. A great deal of confusion surrounds the concepts of simplicity and complexity in contemporary scientific literature. Some investigators hold that simplicity and complexity are merely in the eyes of the beholder, that they are a feature of our descriptions and explanations of empirical phenomena, and not of the phenomena themselves. This is not the view taken here. Simplicity and complexity are considered a measure of the organizational structure of a real-world system. The measure provides information essential to our understanding of the nature and evolution of reality. It can be rendered operational by counting the number of parts in the system together with the connections between them, or by counting the number of "bits" of information that went into the construction of the system. The sciences of complexity are well named: not because they are more complex than their classical predecessors (indeed, they excel in the search for simplicity and elegance in theory formulation and explanation), but because they deal with complex systems, and complex hierarchies of systems.

4. Free energy in a system is inversely related to entropy, as given by the equation $F = E - TS$ (where F is free energy, E is total energy, T is absolute temperature, and S is entropy). At any given temperature,

the smaller the system's entropy, the greater its free energy, and vice versa.

5. There are two varieties of catalytic cycles: cycles of autocatalysis, in which a product of a reaction catalyzes its own synthesis, and cycles of cross-catalysis, in which two different products (or groups of products) catalyze each other's synthesis. An example of autocatalysis is the reactions scheme $X + Y \quad 2X$. Starting from one molecule of X and one of Y, two molecules of X are catalyzed. The chemical rate equation for this reaction is $dX/dt = kXY$. When Y is held at a constant concentration there is an exponential growth in X. Cross-catalytic reaction cycles have been studied in detail by the school of Ilya Prigogine. A model of such reactions, known as the Brusselator, consists of the following steps:

$$
\begin{aligned}
(1) &\quad A \rightarrow X \\
(2) &\quad B + X \rightarrow Y + D \\
(3) &\quad 2X + Y \rightarrow 3X \\
(4) &\quad X \rightarrow E
\end{aligned}
$$

In this reaction model X and Y are intermediate molecules within an overall sequence through which A and B become D and E. In step (2) Y is synthesized from X and B, while in step (3) an additional X is produced through collisions of $2X$ and Y. Thus, while (3) in itself constitutes autocatalysis, (2) and (3) in combination make for cross-catalysis.

6. The empirical evidence for this process is indisputable. Diverse atomic elements converge in molecular aggregates: specific molecules converge in crystals and organic macromolecules; the latter converge in cells and the subcellular building blocks of life; single-celled organisms converge in multi-cellular species; and species of the widest variety converge in ecologies. As each level of organization is attained, evolution brings forth increasingly complex systems on that level. Atomic structures build in time from hydrogen to uranium and beyond; simple chemical molecules give rise to more complex polymers: organic species evolve from unicellular to multicellular forms, and immature ecologies build toward the mature climactic form. Continental ecosystems, which integrate vast and diverse populations in embracing energy-processing cycles, constitute the highest level (but by no means the most complex form) of organization in the biosphere.

7. The standard autopoietic model consists of five basic components: a "hole" (meaning an empty position in the grid of system interaction), a substrate, a simple link, a singly or fully bonded link, and a catalyst. Three basic transformations can be produced by these components: a catalyst and two substrate units can produce a link with a hole as a byproduct and leaving the catalyst unchanged: a free or bonded link can disintegrate into two substrate units and fill available holes; and various bonded links can be produced (a free link can be bonded with a bonded chain, two bonded chains can be bonded into one, or two free links can bond into a chain formation).

8. The theory states that if a process is controlled by some functional relationships (maximizing or minimizing functions), then the outcome of the changes can be explained by specific kinds of catastrophes. The simplest catastrophe, the fold, has one control dimension and one behavior dimension, while a more complex catastrophe, the parabolic, has four control dimensions and two behavior dimensions.

Part 2

The Realms of Evolution

Introduction

General evolution theories remain products of the intellect, fascinating but abstract possibilities unless the processes and products they describe can be tested and verified in the empirical realms of observation. Not everything that is rational is also real [*pace* Hegel]; a general theory may be coherent and intellectually pleasing yet unexemplified by, and thus inapplicable to, empirical reality. The proof of the pudding is in the eating: we must turn to the real world and its realms of evolution to test the validity of the theory outlined in Part One.

The realms of evolution in the empirical world do not follow classical disciplinary boundaries although they are logically aligned with the great divisions of empirical science: divisions between the physical, the biological and the social sciences. In the perspective of a GET these are not absolute and water-tight divisions but sets of theories based on investigations of third-state systems located at different clusters of organizational levels. The physical sciences, more specifically cosmology and astrophysics, deal with the synthesis of matter-energy systems on the most basic levels of organization, beginning with quarks, hadrons, leptons, electrons and neutrons, and all elementary particles with or without rest mass. The biological sciences investigate more complex entities built of physical and physicochemical matter-energy systems, beginning with the nucleic acids that code the genetic information of living species. The social sciences in turn concern themselves with a still higher cluster of organizational levels, namely with the various social and cultural systems constituted by *Homo sapiens*, a particular kind of living species.

Each of these clusters of levels of organization is built on the cluster of levels below. Since the lower levels had to be already formed before the higher levels could appear, it is not surprising that the clusters emerged successively in time; the three great realms of evolution form not only an organizational, but also a temporal sequence. The synthesis of matter began about fifteen billion years ago, an almost infinitesimal fraction of a second after the creation of the observed universe. Living systems emerged sometime between 4.6 and 3.6 billion years ago on the surface of our planet, although they may have originated earlier, and may continue to emerge later, on other planets. The age of the social and cultural systems created by sapients must be measured in tens of thousands of years; they date roughly from the emergence of our ancestors as a dominant species among the hominids.

The three clusters of organizational levels form a pyramid. At the base there are the physical and physicochemical matter-energy systems synthesized throughout the billions of galaxies that populate the universe. Above it are biological systems on earth, and presumably on a significant number of other planetary surfaces. At the apex of the pyramid are sociocultural systems formed on earth by sapients—and (probably) on other locations by extraterrestrial beings with the required forms of intelligence.

Logically, the investigation of the great realms of evolution must start at the base of the pyramid and at the beginning of the process: with the synthesis of matter in the physical universe.

The processes of cosmic evolution bring forth a variety of matter-energy systems; these systems define the range of constraints and the scope of possibilities within which systems on higher levels can evolve. Systems on the lower level clusters can permit the evolution, but can never determine the nature, of systems on higher-level clusters. As already noted, the laws of evolution are not deterministic but possibilistic; they do not select precise

Introduction

evolutionary trajectories but set the context within which nonequilibrium systems in the third state choose their own evolutionary destinies. The evolution of physical matter-energy systems sets the stage and specifies the rules of the game for the evolution of biological species, and biological evolution sets the stage and specifies the rules of the game for the evolution of sociocultural systems.

This exploration of the general theory sketched in Part One will follow the pattern traced by evolution itself: it will start with a review of the evolution of matter in the cosmos, continue with the evolution of life in the biosphere and of societies in history, and conclude with the evolution of mind in the human individual.

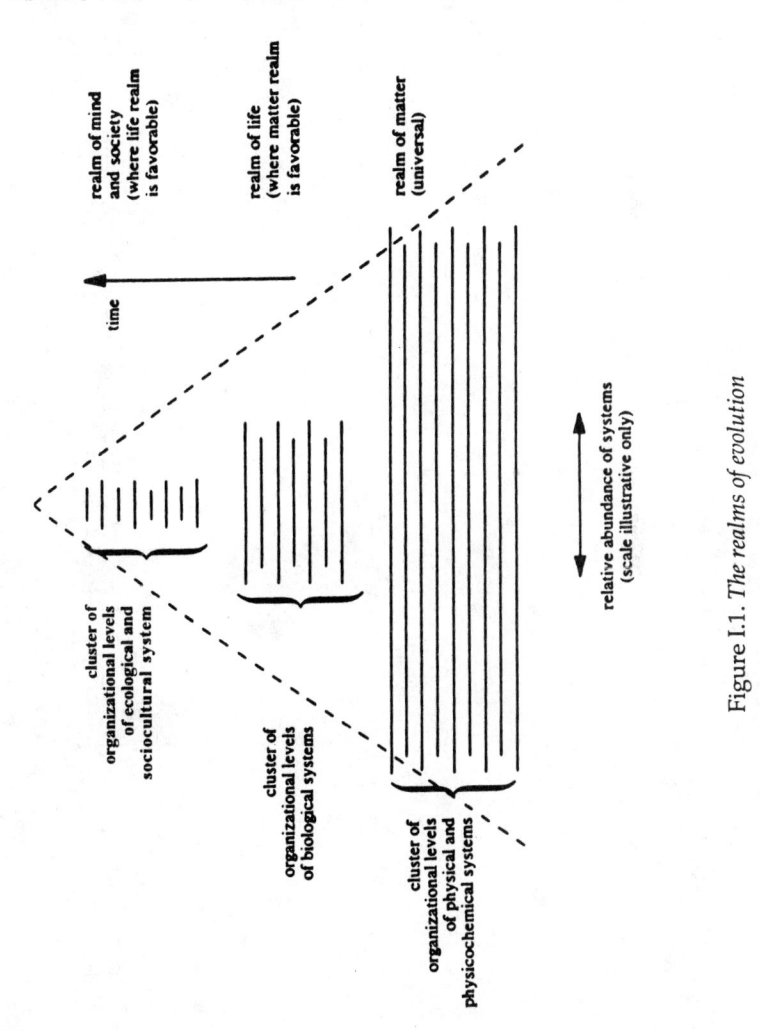

Figure I.1. *The realms of evolution*

3

The Evolution of Matter

We begin the exploration of our general evolution theory by going back to the scientific equivalent of Adam and Eve: to the beginnings of the evolution of matter in the universe. This is the logical starting point: the beginning of time and the beginning of space as we know them, the first and necessarily theoretical microseconds that marked the initiation of all evolutionary processes in our world.

THE ORIGINS OF THE UNIVERSE

The origins of the universe were the focus of philosophical and poetic speculation from the earliest myths to this day. Cosmology moved from the field of myth and metaphysics to that of science in the twentieth century, but it was only in the 1980s that major breakthroughs occurred in our understanding of the very earliest phases of the universe.

Grand Unified Theories (GUTS)

Contemporary cosmologists postulate new, unified and universal forces in a veritable mathematical *tour de force* to account for the observed features of the universe. According to current theories (the so-called GUTs and super-GUTs—grand and super-grand unified theories) the origins of the universe involve both a major discontinuity and the fragmentation of a fully integrated force field. Although GUTs and super-GUTs are still in development and beset with as yet unresolved puzzles, theoretical cosmologists believe that the universal forces that govern cosmic processes are essentially unitary. Super-GUTs assume that during the first phases of the creation of the universe, a single unitary force governed all interactions. Then, as the universe expanded and cooled, this unitary force broke up into the familiar universal forces of gravity, electromagnetism, and the strong and weak nuclear forces.

Contemporary GUTs hope to achieve what Einstein tried to do four decades earlier: create a unified field theory. Einstein attempted to unify what he believed were the two fundamental forces in nature: gravity and electromagnetism. However, the strong and weak nuclear forces were discovered at about the same time, and by midcentury physicists were confronted not with two, but with four seemingly distinct forces.

GUTs became a respected scientific endeavor when the electromagnetic and the weak nuclear forces could be unified in the theory of the *electroweak* force. According to this theory, advanced by Sheldon Glashow, Steven Weinberg and Abdus Salam in the late 1960s and confirmed only a decade later, at high temperatures these forces are one and the same. It is only at lower temperatures that the four bosons that carry the unitary electroweak force (the photon, the W^+, the W^-, and the Z^0) divide into the photon (carrying the electromagnetic force) and the other three (which carry the weak nuclear force). The particles were discovered in 1983, at the CERN accelerator in Geneva. Spurred by this success, physicists are intent on moving toward "grand unification" by integrating the strong nuclear force with the electroweak force in the *electronuclear force*. This version of GUT calls for a finite rest mass in the *e*, *mu*, and *tau* neutrinos—they could then transform into each other. But the theory is not easy to verify: this level of unification occurs at extremely high energies. Proof may elude scientists in the laboratory, but could still be found in the particles left over from the Big Bang—the only event in the known universe that produced the required energies.

Theoretical physicists are now working on the final step known as supergrand unification: they seek to demonstrate the unity of the electronuclear force with the force of gravity. The mathematics of these "theories of everything" (TOEs) are often outlandish. In the theory of superstrings, for example, the computations call for a universe of ten dimensions—physicists hope to reduce them eventually to four. Although TOEs are still speculative, most physicists agree that the derivation of a *grand unified force* requires the decay of the superheavy X-bosons within the first 10^{-35} second, owing to falling temperatures.

A *super-grand unified force* is assumed to have operated at temperatures of 10^{30} K, reigning until about 10^{-39} second in the current phase of the universe. When time advanced to 10^{-35} second, the temperature dropped to 10^{28} K and a phase change took place in the expanding cosmos: below this temperature X-bosons could no longer be produced. As these bosons vanished the expanding cosmos developed a tension that, according to the GUTs, caused a rapid acceleration in expansion, that is, created the so-called inflationary phase. At the conclusion of inflation the X-bosons disappeared and the grand unified force fragmented into the electroweak and the strong nuclear forces. The cosmos began to expand at a more moderate rate.[1]

Inflationary Scenarios

The hypothesis of an inflationary phase is to account for the observed isotropy and homogeneity of the universe. The distribution of photons, for example, is isotropic (that is, the same in any direction) within a limit of accuracy of one part in ten thousand. Yet only those regions that are less than two degrees apart could have been in contact with each other when photons decoupled from radiation. Thus the 2.7 K background radiation, filling space evenly, seems a puzzle. To account for this "horizon problem" and related problems in Big Bang cosmology, Alan Guth proposed the inflationary scenario. The first formulation of the theory, in 1981, proved to be fatally flawed; but the "new inflationary scenarios" advanced since 1983 by the Russian physicist A. D. Linde, among others, appear to better conform to observational evidence.

The flatness, homogeneity, isotropy, and baryon asymmetry of the universe, the existence of galaxies, and the absence of monopoles and domain walls are explained in the new scenarios. They offer an especially ingenious solution to the problem of isotropic background radiation. Before inflation the region of the primordial universe that inflated into the known universe is conceived to have been small enough to produce thermal equilibrium. Employing complex and admittedly speculative theoretical devices (such as a special type of Higgs potential to allow a smooth phase transition through a large-scale supercooking effect), the new inflationary scenarios postulate a chaotic phase of random fluctuation following the initial instability, followed by the rapid inflation of one of the fluctuation "bubbles" into the known universe. At the same time long-wave quantum fluctuations of curvature and of the scalar field are believed to have grown rapidly, introducing the density perturbation necessary for the subsequent formation of galaxies. (An entirely "wrinkle-free" early universe could not have produced the inhomogeneities that subsequently led to the gravitational contraction of matter into galaxies.)

According to the inflationary scenarios the universe underwent a phase change prior to settling into the kind of order we observe today. The phase change consisted of the following sequence of states:

1. An initial state of net zero energy (a "vacuum") during which neither time nor space were measurable in relation to time and space in the subsequent epochs.
2. A transitory chaotic state of an indefinite (and immeasurable) duration, consisting of random fluctuations and "bubbles" (induced possibly by a critical instability of the initial state)
3. A short-lived inflationary state, lasting 10^{-33} second during which one of the bubbles nucleated and inflated by a factor of 10^{20} to 10^{50}.
4. A consequent dynamically ordered state lasting to date some 10^{18} seconds (about 15 billion years) in which the known universe thins and cools and creates third-state matter-energy systems progressively further from thermodynamic equilibrium.[2]

THE SYNTHESIS OF MATTER-ENERGY SYSTEMS

To explain the generation of matter in cosmic space-time is a basic aim of contemporary cosmology. Cosmologists now assume that the energy of the vacuum from which the known universe originated has split and transformed into positive energy, which is matter, and negative energy, which is the energy of the gravitational field. When the energy of the field is transferred into matter energy, material systems come into being. An irreversibility is introduced into the structures of the universe that did not exist in the standard model produced by Einstein in 1917. However, the solutions of Einstein's equations were shown to suggest an unstable universe by Friedmann in 1922; Hubble's observations confirmed the recession of the galaxies; and the discovery of residual black-body radiation testified to the thermal evolution of the cosmos. Irreversible processes began to assert themselves in the forefront of cosmological speculations.

In a new hypothesis, Prigogine and Géhéniau attempt to reconcile the concept of an irreversibly evolving universe with Einstein's field equations. They postulate a new equivalence, complementing relativity theory's equivalence of matter and energy: this is the equivalence between matter and entropy. If matter is created from the vacuum, the entropy of the latter must have been zero: when matter is created, entropy is also produced. The entropy necessary to generate matter proves to be gravitational entropy.[3] If the new hypothesis is valid, the universe is self-creative in its very nature. Evolution in the form of the irreversible creation of matter becomes one of its fundamental characteristics.

The matter-content of the contemporary cosmos is but a fraction of the matter created in the first fractions of a second following the Big Bang. The observed universe consists entirely of that segment of matter that survived the very early phases in virtue of a slight imbalance between matter and antimatter. The decay of the X-bosons appears to have yielded a surplus of one proton for every billion pairs of protons and antiprotons. This residue (one-billionth part of the originally created matter of the universe) survived the pair-annihilation process and became available for further evolution in cosmic space and time.

Temperature and Density Parameters

The ongoing evolution of matter in the cosmos represents the transfer of energy from the cosmic radiation field, governed by the action of universal forces and the changing parameters of the expanding universe. Two factors of change are of special importance: the temperature and the density of the cosmos.

When the inflationary phase ended, the density and the temperature of the cosmic fireball were too high to permit the formation of systems of any kind. Both these values were in a range that is humanly unimaginable. A universe younger than 10^{-24} second had densities greater than 10^{50} grams per cubic centimeter [g/cm^3] and temperatures in excess of 10^{20} degrees on the Kelvin scale.

However, this intense and dense sea of radiation soon provided a milieu for the emergence of the first microparticles. At already <10^{-11} second quarks must have existed, although details of their creation are still lacking. At about 10^{-11} second and 10^{15} K temperature, hadrons (protons, neutrons, and mesons) are believed to have formed. They were created by "pair-creation"—the opposite of the process of pair-annihilation. (In pair-creation, particles emerge from clashes among packets of energy, in a way opposite to that in which colliding particles and antiparticles—protons and antiprotons—annihilate each other and yield pure energy.) As density and temperature decreased further, fewer and fewer hadrons were created; the dominant process became the breakdown of the existing hadrons into high-energy photons.

After about a millisecond the superenergetic conditions required for hadron-creation and annihilation had almost subsided, and lighter particles, called leptons (electrons, neutrinos, and muons) began to predominate. A whole class of elementary particles was created under conditions where the average temperature had decreased to "merely" 10^{10} K and the average density was around 10^{10} g/cm^3. By the completion of the first millisecond, most of the leptons, too, had broken down into photons: the cosmos was still too hot and dense to permit the formation of enduring matter energy systems. The matter content of the cosmos had thinned once more: whatever particles survived this epoch did so as a microscopic precipitate within the macroscopic sea of radiation that still filled space and time.

The universe continued to cool and thin, and allowed the surviving particles to become electromagnetically linked. Hydrogen was the first element to be created, consisting of the linkage of but one negatively charged electron with one positively charged proton. When temperatures dropped to about 10^9 K, protons and neutrons interacted to produce deuterium, an isotope of hydrogen. As deuterium interacted in turn to produce helium,

THE EVOLUTION OF MATTER

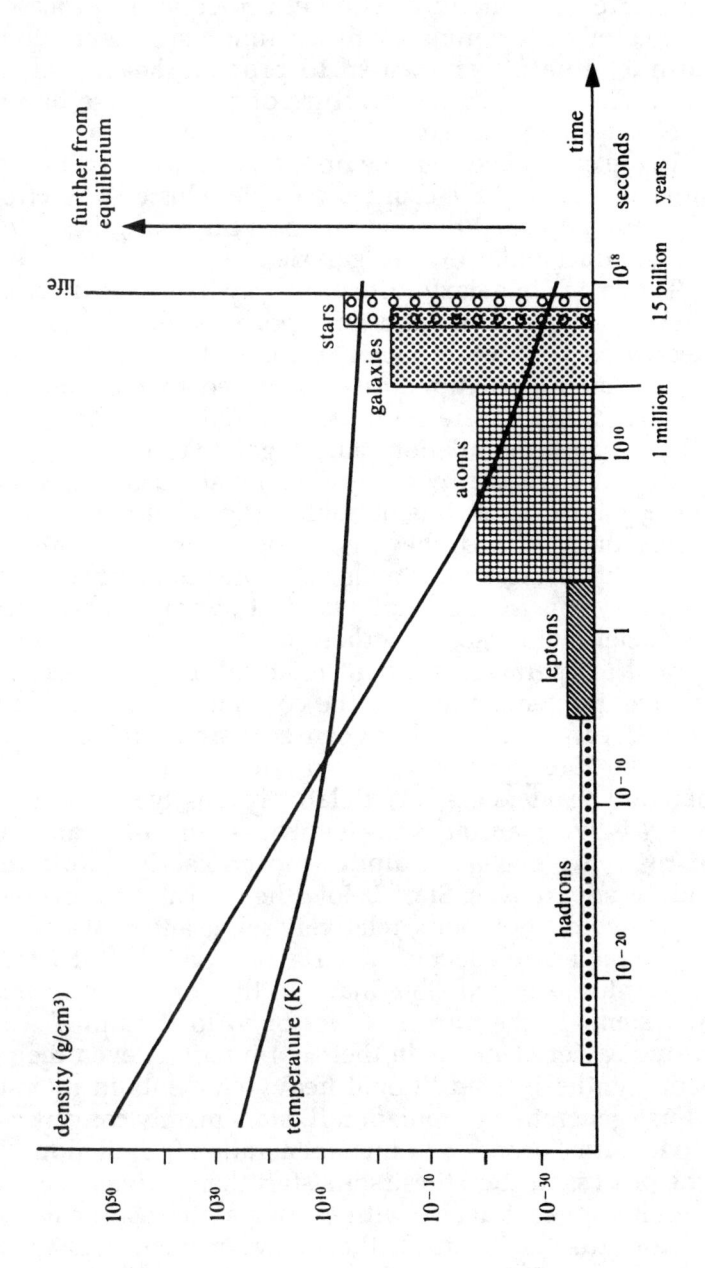

Figure 3.1. Density and temperature parameters in the buildup of matter-energy systems

some 20 to 25 percent of the hydrogen component of the universe was converted into helium. The helium-creating process may have lasted only a few minutes. By the time a sufficient number of helium nuclei had been created to produce heavier elements through interaction, the temperature of the universe had fallen below the threshold value of 10^8 K.

In the epoch between the first few thousand years and one million years, most of the charged particles clustered electromagnetically into atoms. At the one-million-year mark, these systems began to contract and form vast galaxies.[4]

The epoch of galaxy formation may have lasted about five billion years. Midway through this epoch the density of the cosmos decreased by another factor of more than a billion, to 10^{-20} g/cm^3, and average temperatures dropped to about 300 K. As soon as the galaxies were essentially complete, perhaps already one billion years after inflation, stars began to form.

Stars are created by the gravitational clustering of matter within the galaxies. But stars not only form within the vast galactic reaches; once formed; they also evolve. First-generation stars begin their life cycle with the already synthesized matter of the universe: they consist of about 90 percent hydrogen and 9 percent helium nuclei, with traces of other elements. Initially they have but a small temperature gradient relative to the surface. But as stars evolve, the temperature of the core increases and therewith the temperature gradient between core and surface. Nuclear fusion in the core synthesizes the hydrogen nuclei into helium, and then into heavier elements. Relatively massive stars—equal to or more massive than our sun—evolve significantly and, at the end of their life cycles, explode (more exactly, implode and rebound) as supernovae. Stars below the critical mass evolve less radically, toward a luminous, relatively stable and cool state.

The remarkable fact of stellar evolution is that it sets forth, in a constant and irreversible manner, the formation of matter-energy systems in the universe. Not only do stars produce progressively heavier elements in their hot interiors, even their final explosion synthesizes additional heavy elements in interstellar space. First-generation (population II) stars mainly convert hydrogen into helium. Second- and third-generation (population I) stars carry the process further: these stars start their active life cycle not only with hydrogen, but also with elements such as iron and magnesium, and they build up further heavy elements. The synthesized elements ultimately find their way into interstellar space. Even the cooler and less massive stars inject the elements synthesized in their interiors into surrounding space; they continually lose their outer layers in low-velocity stellar winds and occasionally eject deepers layers in bursts of activity that produce planetary nebulae.

The statistically irreversible buildup of matter in the cosmos results in a rich soup of elements that undergoes further evolution in successive generations of stars and enriches vast interstellar clouds. The more dense among these clouds are known to contain heavy substances including various organic molecules such as formaldehyde and hydrogen cyanide.[5]

In today's relatively cool (2.7 K) and thin (approximately 10^{-30} g/cm^3) universe, the synthesis of matter-energy systems continues. Progressively heavier elements are produced, including oxygen, nitrogen, and carbon—elements essential for the evolution of life. As hydrogen is present as well, it is probable that in many places the universe harbors matter-energy systems we would call living. Wherever suitable energy and density conditions are given—that is, on the surface of planets similar to earth associated with a star similar to our sun—the evolution of complex molecular systems is likely to begin. The thermal and chemical conditions for this process are known: there must be a variety of atoms capable of forming chemical bonds (mainly carbon, hydrogen, nitrogen, oxygen, and also phosphate and sulphur supplemented with some metals and halogens); the temperature must not exceed 100°C; and there must be a continuous energy irradiation. With about 200 billion stars in our galaxy and some 10 billion galaxies in the universe, chances are good that a significant number of stars have planets that meet these specifications. Since the abundance and distribution of chemical elements are much the same in all parts of the universe, the probability that there are a considerable number of planets with matter-energy systems that qualify for the predicate "living" is by no means negligible.[6]

The universe is at the same time more complex and more unitary than we had reason to believe until very recently. It consists of a radiation field of enormously high initial energy which cools, expands, and brings forth a great variety of nonequilibrium matter-energy systems. (Neutrons, protons, electrons, in fact all particles with a positive rest mass are already in a state of nonequilibrium; only photons with zero rest mass are an exception.) As field energy is transferred, and due to expansion density and temperature decrease, an increasing variety of material particles are created. This is followed by the synthesis of more complex structures still further from thermodynamic equilibrium, such as atomic nuclei.

Gravity condenses these matter-energy systems into galaxies, which subsequently produce radiant stars. Within the hot interior of stars, heavier elements are constantly synthesized, and the supernovae that mark the end of massive stars produce the temperatures required for the synthesis of still heavier elements. The building of matter-energy systems continues with the gravitational contraction of already enriched interstellar matter and the formation of second- and third-generation stars. Stellar and interstellar matter becomes sufficiently enriched to initiate the synthesis of nonequilibrium systems in the third state on planetary surfaces with a suitable chemical composition exposed to a constant energy flux within the permissible temperature range.

THE ARROW OF TIME IN THE COSMOS

The Radiation-to-Matter Energy Transfer

On the assumption that the energy of the vacuum from which the cosmos originated has split into a positive segment which is matter and a negative segment which is the gravitational field, the creation of matter involves the transfer of the energy of the field to the energy of matter. Given that the universe expands and cools, the radiation which was predominant in the field diminishes while matter is continuously built up. In the expanding and cooling universe *energy is irreversibly transferred from radiation to matter*. The arrow of time in the cosmos is given by this irreversibility. As energy is more and more concentrated in matter-energy systems, these systems move into the third state, further and further from thermodynamic equilibrium.

The transfer of energy from radiation to matter can be calculated with mathematical precision. The energy transferred is expressed as energy density, that is, as unit of energy per unit of volume. First the energy density of the background radiation is calculated, and then it is shown how this relatively low quantity (equivalent to 2.7 K) changes over time. The formula allows the extrapolation of a curve that traces the decrease of the energy density contained in pure radiation from the first fraction of a second to any subsequent time in the history of the universe.

Next the energy density retained in matter is computed, and its change over time is plotted. Since energy is constant while the universe is expanding, the density of energy in the cosmos (whether unbound in radiation or retained in matter-energy systems) decreases over time. But, because radiation, unlike matter, is affected linearly by the Doppler shift (that is, by the recession of the galaxies), the energy density unbound in pure radiation decreases faster than the energy density bound in matter.[7]

The difference in the decrease of energy density in pure radiation ($\rho_r c^2$) and in matter ($\rho_n c^2$) makes for a net and continuous transfer from radiation to matter. The process can be plotted in the form of two curves; these intersect sometime between 50,000 and one million years after the initial inflationary phase of the universe. (The variance is due to uncertainties in determining the precise amount of the energy density of matter.)

Viewed in these terms, the course of cosmic evolution becomes logical and clear. During the first microseconds, the super-unified field of the universe consisted almost entirely of radiation. As the universe cooled and thinned, matter-energy systems came to be synthesized and energy became increasingly bound in matter. When the energy-density of radiation and of matter matched, matter decoupled from radiation: photons and free electrons could no longer interact. The universe as a whole moved into thermal disequilibrium. It created flows of energy between energy-radiating macrosystems, such as active stars, and various types of microsystems engulfed in the flows in interstellar clouds and on the surface of planets. Some among the many energy flows proved to be the engine of evolution for matter-energy microsystems of ever increasing dynamism and complexity.

The differential between the energy density of matter and of radiation increases over time. The structural complexity and dynamism—in other words, the negative entropy—of the microsystems grow proportionately to this differential. When these systems reach a critical phase of negentropy, they no longer behave as systems in the first state *in* equilibrium, or as systems in the second stage *near* equilibrium. but as systems in the third state *far* from the equilibrium state. Immersed in an energy flux, they do not go linearly to equilibrium but maintain structure and complexity in dynamic steady states. They may also increase their level of nonequilibrium by moving into alternative, still more structured steady states following critical perturbations.

As third-state systems move further into nonequilibrium, the density of the energy flux captured and retained by them increases. While the average energy density of matter in the cosmos decreases (although slower than the energy density of radia-

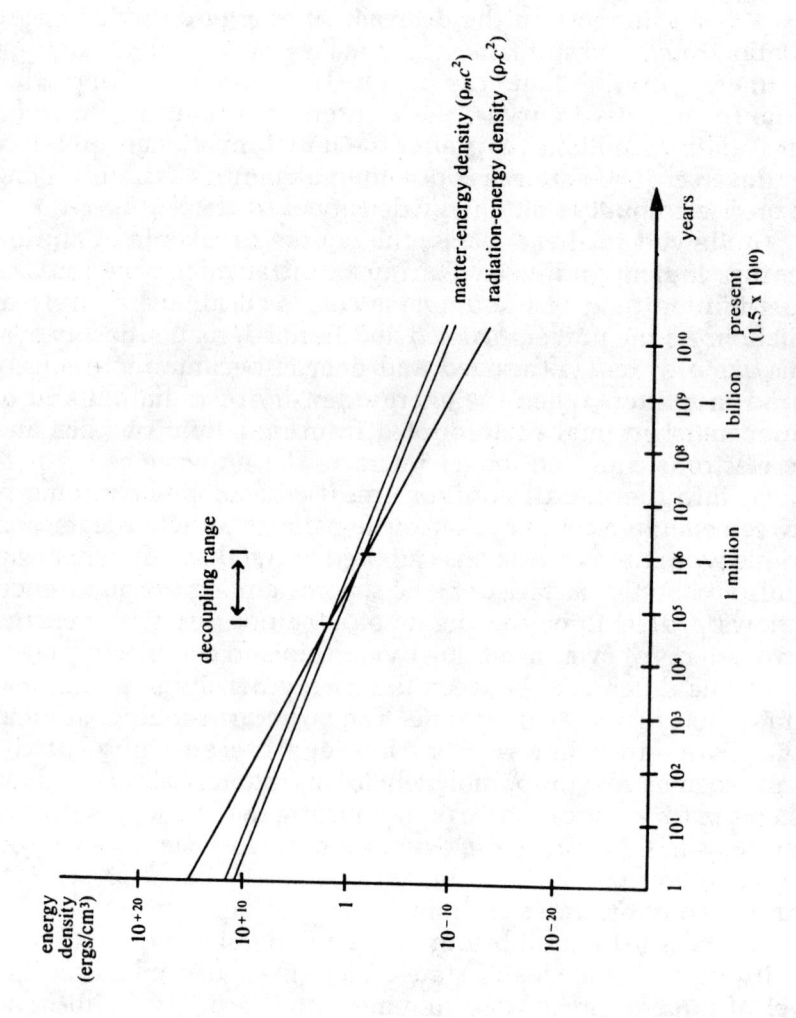

Figure 3.2. *The decrease of energy density in radiation and in matter*

tion), the net energy transfer from radiation to matter creates energy flows that trigger the evolution of dynamic systems that retain an increasing amount of free energy per unit of mass.

The evolution of matter in the universe is thus assured. Out of a disequilibrated cosmic radiation field come matter-energy systems in the third state, more and more negentropic, complex and dynamic, and removed from thermal and chemical equilibrium.

The synthesis of dynamic matter-energy systems that explore ever further reaches of nonequilibrium continues today, some 15 billion years after the original instability that marked the beginning of space and time as we know them. While no new galaxies are being formed, within those that exist new generations of stars are born and massive older stars implode and fly apart in supernovae or collapse into black holes. Stellar and interstellar matter is irreversibly enriched with heavier elements. The stage is thus constantly prepared for the emergence of the still more structured matter- energy systems that we associate with the phenomenon of life.[8]

The progressive structuring of systems is likely to occur throughout the reaches of the cosmos; the process is triggered in the energy flows created by the progressive transfer of energy from field to matter. While this creates the basic preconditions for the evolution of matter, it does not determine just *how* matter evolves—its precise evolutionary trajectory. Even if the thermal and chemical evolution of stars is likely to be basically similar in all galaxies, significant differences could occur in the trajectories traced by more complex nonequilibrium microsystems. Elsewhere in the cosmos macromolecules, cellular, supracellular, ecological and possibly even sociocultural systems may emerge as well, but they may assume forms considerably different from those which they take on earth.

Notes

1. Super-GUTs attempt to integrate all forces and interactions observed in the universe. They unify leptons and quarks, believed to be the basic elements of matter in the cosmos, by a gauge symmetry valid at very small distances (less than 10^{-16}cm). At larger distances the differences between quarks and leptons are assigned to a breaking of

the symmetry. Two kinds of gauge symmetry are put forward to account for the unity of quarks and leptons. One, described by the group SU(5), consists of a family of fifteen fermions divided into two classes, of ten and five fermions respectively. The other symmetry, known as SO(10), is more stringent in that it relates all sixteen fermions to each other. Moreover, the parameters of the symmetry can be adjusted to match observable quantities. Another symmetry, relating fermions and bosons, turns out to be a form of supersymmetry: it postulates supersymmetric (and extremely heavy) "partners" to the observed leptons, quarks, and bosons. However, super-GUTs are as yet speculative. Fully developing the theory would mean integrating quantum theory into relativity theory, i.e., creating a quantum theory of gravity. This is a challenge that still awaits theoretical physicists.

2. It is interesting to note that this phase change corresponds remarkably to the process of bifurcation in nonequilibrium systems, i.e., to a basic feature of evolution within the cosmos. The sequence of states adds up to the phase portrait of an evolutionary bifurcation. (Compare with Figure 2.9 in Chapter 2.) There is an initial steady state which is destabilized and gives rise to a phase of chaos and indeterminacy; there is then a sudden leap into a new state through the amplification of some element in the randomly fluctuating system; and finally there is the emergence of a new and relatively stable dynamic regime which proves to be the initial state of a phase of progressive system building or self-complexification. The inflationary phase, like all processes of catastrophic bifurcation in evolving systems, separates the consequent phase from the initial phase: the nature of the observed order in the cosmos is not a function of whatever order may have reigned prior to inflation.

The interplay of determinism/indeterminism, and high/low entropy production correspond as well. There is a remarkable degree of order and determinism in the consequent phase, while chaos and indeterminism reign in the transitory chaotic phase. Entropy production reaches a maximum at the end of the inflationary state, as in all phases of bifurcation, and decreases as the new dynamic regime is established. If the evident similarity between the still speculative description of the very early universe and catastrophic bifurcations in nonequilibrium systems is not a superficial analogy or mere coincidence but a basic invariance at the root of reality, the processes of evolution *in* the cosmos repeat, in fact recapitulate, the evolutionary dynamic *of* the cosmos.

3. Prigogine and Géhéniau use the Landau-Lifshitz pseudotensors together with conformal (Minkowski) coordinates in their attempt to reconcile the evolution of the universe with the Einsteinian model. As this method suffers from the defect that, as a result of using pseudotensors, it is not covariant, Prigogine and Géhéniau propose a supplementary "C field" corresponding to the creation of matter in Einstein's field equation. The role of the C field is to express the cou-

pling between gravitation and matter leading to the irreversible production of entropy. See I. Prigogine and J. Géhéniau, "Entropy, Matter, and Cosmology," *Proc. National Academy of Science* (U.S.A.) 83 (1986).

4. The exact time when galaxies began to form is still controversial: standard cosmological theories yield a figure for galactic mass that is far smaller than that indicated by observation. Some astrophysicists suggest that the universe may contain much "dark matter" (invisible to optical telescopes but registered by radio astronomy). Such matter could have influenced the formation of galaxies and could account for the noted discrepancy.

5. More than fifty interstellar molecules have been identified so far, many with molecular weights exceeding fifty atomic mass units. Astrophysicists have also observed nearly two hundred additional unidentified features in the relevant millimeter-wave spectrum, while new detection techniques revealed the presence of amino acids as well as nucleic acids in meteorites.

6. Astronomer Frank Drake proposed an equation to calculate the number of civilizations in our galaxy that have the technology and the motivation to communicate with us. It is $N = R^* f^p n^e f^l f^i f^c L$. Here R equals the rate at which stars form. This is multiplied by the fraction of stars with planets, times the average number of planets physically capable of supporting life, times the fraction of the planets that actually give rise to life, times the fraction of life-supporting planets that generate intelligence, times the fraction of intelligent species that have the technology and the motivation to communicate, times the average life expectancy of civilizations that attempt communication. If we are interested simply in the occurrence of life, the probabilities increase sharply. Assuming the nonaccidental emergence of life (i.e., that all planets physically capable of supporting life actually give rise to life), the indicated equation is $N = R^* f^p n^e$: the rate of star formation times the fraction of stars with planets times the average number of planets that can physically support life.

7. The relevant equations are the following:
   (1) Cosmic background radiation:

Temperature	$T = 2.7 \pm 0.1 K$;
Energy density	$u = a(2.7)^4 \sim 10^{-13}$ ergs cm^{-3}
where	a is the "radiation constant":
	7.6×10^{-15} erg $cm^{-3} K^{-4}$.

   (2) As $u \propto T^4$.
   the change of radiation temperature with time is:
   $$T = 2.7(1 + z) \cong 10^{10} t^{-1/2}$$
   where $z = v/c$,
   v is recessional velocity,
   c is speed of light,
   and t is given in seconds.

(3) Present 'equivalent' energy density of matter
   $\rho_m c^2 \sim 10^{-8}$ erg cm^{-3}.
(4) The temporal variance of the density of matter:
   $\rho_m \sim 10^6\ t^{-2}$.
(5) The average temperature of matter at any time t:
   $T \sim 6 \times 10^{16}\ t^{-1}$,
(6) The average temperature of radiation at any time t:
   $T \sim 10^{10}\ t^{-1/2}$,
(7) The energy density of radiation [from (2)]:
   aT^4.
(8) The energy density of matter [from (3)]:
   $\rho_m c^2$;
(9) Temporal change in the proportion of energy density in matter and in radiation:
   when $t < 500{,}000$ years: $\rho_m c^2 < aT^4$
   when $t = 500{,}000$ years: $\rho_m c^2 = aT^4$
   when $t > 500.000$ years: $\rho_m c^2 > aT^4$

For details see the appendix to Eric Chaisson, *The Life Era* (Boston: Atlantic Monthly Press, 1987).

8. The cosmic evolution of matter stretches to an infinite time horizon if the universe is open—that is, if the mass of matter created in the course of time does not exceed the critical value of approximately 10^{-30} gram/cm^3. Above this value the universe will contract back and end in a terminal Big Crunch; below it it will expand forever. Current measurements are not yet sufficiently refined to decide the exact value of matter density in the cosmos, hence we still cannot tell whether we live in an open or in a closed universe.

4

The Evolution of Life

The evolution of life on earth began with initial conditions provided by evolution in the cosmos. That cosmic evolution created precisely those particles, atoms, and molecules that it did determined the range of possibilities and the parallel constraints for the evolution of biological species, but it did not determine their exact forms. Living systems define their own evolutionary trajectories. They are neither determined by, nor reducible to, physical and chemical systems and processes.

Biological evolution is not merely cosmic evolution on a planet: it would be naive to search for products and processes that are identical in the two realms. Yet the general laws that govern the evolution of all systems in the third state remain invariant as we pass from the cluster of organizational levels prevalent in the universe at large to the cluster occupied by living systems in the earth's biosphere. We can explore this invariance by reviewing recent discoveries concerning the origin of life, the nature of macroevolution and of mutation, and the emergence of *Homo sapiens*.

THE ORIGINS OF BIOLOGIC EVOLUTION

According to current estimates, biological evolution on this planet began over 3.6 billion years ago. Fossils with traces of advanced chemical evolution dating back 3.4 billion years have been successfully identified; primitive biological organisms have been shown to have existed at least 2.8 billion years ago, and evidence of the biochemical activity of prokaryotic cells with modern enzyme structures has been uncovered in fossils 2.3 billion years old.

Statistical studies show that the probability of the accidental appearance of self-reproducing systems of linear polymers is extremely small; at the same time, experimental evidence indicates that whenever certain chemical and energetic conditions obtain systems similar to the primitive forms of life are synthesized. Most scientists now accept the view that under the particular conditions that reigned on earth some 4.6 to 3.6 billion years ago, life arose necessarily rather than as a result of some cosmic accident.

Theories of a purely accidental (not to mention extraterrestrial) origin of life, popular in recent years, are unduly speculative. The chemical constituents required for the evolution of life were present on earth before biological evolution took off. The six elements that make up some 98 percent of matter in the cosmos—hydrogen, helium, carbon, nitrogen, oxygen, and neon—and even the more complex molecules essential for the synthesis of the first self-replicating cells, were already synthesized in the physical universe. Even amino acids and nucleic acids were (and are) synthesized; they are known to be present in meteorites. We now know that the universe itself produces the basic building blocks of life, and that some of these building blocks could have found their way to the surface of the youthful earth, carried most likely by grains of dust from dense interstellar clouds.

Conditions on earth were ideal for the abiogenic synthesis of the chemical constituents of life. Monomers like sugars, amino acids, purine, and pyrimidine bases, and linear polymers built of these monomers, such as proteins, nucleic acids and other macromolecules, must have been synthesized in the constant energy flow from the sun and from the hot magma of the earth's core. The chemical composition of the planetary surface was a thermodynamically mature template capable of initiating biological processes. It absorbed solar and geothermal energy, stored and used some—initially minuscule—part of it in matter-energy structures, and radiated the rest into space.

The constant flow of energy from the sun to earth, and from the interior of the earth to its surface, must have had the effect of organizing the elements of the "chemical soup" on the surface into protobionts: systems capable of capturing and storing ever more of the free-energy flux. Experiments show that in a system containing the main bioelements (carbon, oxygen, hydrogen, nitrogen, phosphorus, and sulfur), the energy of visible and ultraviolet light triggers a series of photochemical reactions the end result of which are various compounds of medium molecular weight.

Experiments that seek to reproduce the origins of life in the laboratory have chalked up an impressive record since the pioneering efforts of A.I. Oparin in the 1930s. By the 1950s Stanley Miller and Howard Urey conducted path-breaking studies in which they exposed a mixture of water vapor, methane, ammonia, and hydrogen gases to electric discharges. They found that after a week organic compounds emerged, including D and L amino acids. The nucleotides which make up DNA and RNA molecules eluded detection until the early 1980s, but then C. Ponnamperuma used improved detection techniques and found that, together with the amino acids, all the bases that form DNA and RNA molecules are synthesized as well. Experimenters note that the forces of intermolecular attraction and the tendency of certain molecules to form liquid crystals leads to the spontaneous synthesis of large and complex molecules. Such reactions are known to take place in free solution in water, as well as when the molecules adhere to particles of clay. Indeed, the presence of clay orients the reactions in a direction consistent with the evolution of life.

An intriguing hypothesis concerning the origins of life is not the result of laboratory experiments but of theory combined with observation. It is the work of John Corliss and his co-workers who, after detailed submarine researches, suggest that life originated in shallow Archean seas owing to the eruption of magma through cracks in the earth's crust. Biological evolution on this planet took off, it appears, with the selection of the most rapidly growing and replicating primitive cells within submarine hot springs that acted as small steady-state flow reactors.

Indeed, submarine hot springs are packed-bed, mixing-gradient, continuous-flow chemical reactors. They consist of a cracking front, where sea water is heated rapidly to approximately 600°C, carbon is extracted from the rock, and reactions with ferrous iron produce a reduced fluid containing methane and hydrogen. The fluid reacts and equilibrates with the rock as it mixes with sea water and cools to about 350°C. The equilibrated fluid then either rises directly to the sea floor or mixes in the rock with cold sea water to emerge at lower temperatures (5°-30°C), following a specific, highly constrained mixing-process. High energy monomers could be synthesized in the fluid at or near the cracking front, and then rapidly quenched by mixing. The mixing trajectory would "freeze" the thermal energy into high-energy molecular bonds. Fractures in the upper parts of the hot springs would provide a suitable environment for the synthesis of clay minerals. The presence of clay would extract and accumulate organic matter, achieving high concentrations from an initially dilute aqueous solution. The steady flow within the hot springs would provide a constant supply of energy-rich molecules, including high-energy organic molecules and a variety of oxidized and reduced inorganic molecules and ions.

There is considerable evidence that flow reactors of this kind may be the ideal site for the emergence of dynamic systems in the third state. Katchalsky and Prigogine show that flow reactors constitute classic reaction-diffusion systems basic to the thermodynamic description of self-organizing behavior, while Eigen suggests that they can generate autocatalytic cycles of interacting protein and nucleic acid fragments. Corliss contends that submarine hot springs (which still exist, though in smaller numbers than in the early history of the earth) were the reactors that led to the evolution of the most rapidly growing varieties of cells, that is, those with the most efficient autocatalytic cycles.

In time some of the emerging cells became linked by hypercycles into interacting cohorts. As one species of cell produced material necessary for the growth of another, and the other in turn produced material for the growth of the first—or of a third, which in turn became a link in a chain that fed back to the first—all changes that would destroy one unit in the chain would have destroyed the entire chain. The full chain, however, was richer in information than any of its components and could adapt to conditions that would have proven fatal to the units separately. Thus the evolutionary cycle that took off with catalytic cycles linking protocells and molecules continued with hypercycles linking cells in multicellular systems.

Once evolved, life became almost entirely independent of geothermal energies. Although some marine organisms (giant tube worms, mussels, clams, and crabs) still live in symbiosis with bacteria on geothermal energies at great ocean depths—up to 2,500 meters—life on earth is now sustained by the flow of energy from the sun. Plants use sunlight in photosynthesis, converting water and carbon dioxide into carbohydrates; animals eat plants and other animals. The flow of energy remains, however, an essential engine of all processes in the biosphere. Were the energy differential between the surface of the sun (currently approximately 6000°C) and the surface of the earth (about 25°C) ever to equalize, not only life, but all thermodynamic processes on this planet would soon come to an end. The heat stored in the earth's atmosphere would be depleted in a few months, while that in the oceans would be dissipated already in a matter of weeks. Worms, clams, and bacteria at the bottom of the deepest seas would be the sole survivors.

THE DYNAMICS OF BIOLOGIC EVOLUTION

Speciation

Until the 1980s most life scientists held that the emergence of new species—now known as "speciation"—was essentially correctly grasped in the Darwinian framework, at least in its modern variant known as the synthetic theory. However, an increasing number of biologists are now questioning this conception. Their theories differ from the synthetic theory in several important ways. First, they contest the element of chance that governs the Darwinian account.[1] Researchers find it difficult to see how a basically random search among a vast number of possibilities could have resulted in the emergence of the empirically known complexity of the living world. Michael Denton, for example, asks whether it is credible that random processes could have constructed an evolutionary sequence of which even such basic elements as a protein or a gene are complex beyond human creative capacities. How can one account statistically for the chance emergence of systems of truly great complexity, such as the mammalian brain, when even one percent of the connections in such a brain, if specifically organized, constitute a larger number of connections than the world's entire communications network! Chance mutations acted on by natural selection could well account for variations *within* given species, but hardly for successive variations *among* them.

The standard synthetic theory is in difficulty also when it comes to describing the tempo and mode of evolution. Some paleobiologists attack the classical conception that natural selection, acting on individuals, is gradual and continuous. They claim that "phylogenetic incrementalism," associated with the name of Darwin, is incorrect. They point out that Darwin himself professed incrementalism more from a conservative predisposition than on the basis of scientific evidence. In Chapter 15 of *The Origin of Species*, Darwin declared, "Natural selection . . . can produce no great or sudden modifications; it can act only by short and slow steps." Darwin followed Linnaeus in affirming, "*Natura non facit saltum.*" Sudden leaps in nature resemble discomfortingly sudden changes—even revolutions—in human society, and the dominant mentality of Darwin's time extolled piecemeal adjustments and abhorred wholesale transformations. As current biographical researches show, Darwin was most likely influenced by the dominant mentality of his time. Nature, however, disregards the dispositions of nineteenth-century Englishmen and progresses by sudden leaps and deep-seated transformations rather than through piecemeal adjustments. In 1972, almost one hundred and twenty years after the original publication of *The Origin of Species*, Jay Gould and Niles Eldredge came out with a seminal study ("Punctuated Equilibria: An Alternative to Phylogenetic Gradualism") that introduced the leap into neo-Darwinian biology.

In the theory of punctuated equilibria—where the concept of equilibrium means a dynamic balance between species and environment and not thermal or chemical equilibrium—evolutionary processes concern entire species rather than individual reproducers and survivors. Evolution occurs when the dominant population within a "clade" (a set of species sharing a similar adaptive plan) is destabilized in its milieu and other species or subspecies that emerged haphazardly on the periphery break through the cycles of dominance. At that point the stasis of the epoch is broken, and there is an evolutionary leap from the formerly dominant species, threatened with extinction, to the peripheral species or subspecies. The process is relatively sudden: "speciation"—the emergence of new species—punctuates long periods during which existing species persist basically unchanged and no new species establish themselves.

As long as a species persists, it remains relatively unchanged: its genetic information pool is handed down basically preserved to succeeding generations. Gaps in the fossil record do not indicate a gap in the record of continuous, piecemeal evolutionary changes but signal periods of perhaps many millions of years during which species did not evolve significantly. New species tend to burst on the scene within much shorter time frames: somewhere between 5,000 and 50,000 years. The alternation of long periods of stasis with sudden evolutionary leaps prompted biologist D. V. Ager to liken the history of life on earth to the life of a soldier: it consists of long periods of boredom interspersed with short periods of terror.

The theory of punctuated equilibria is considerably different from traditional Darwinian theory despite the occasional affirmation that it is basically Darwinian evolution applied to entire species rather than to individuals within species. Although the theory does not specifically question the accidental nature of mutations (which may make it inadequate to explain the emergence of vast complexity and higher organizational levels within the known time frame of evolution on earth), the distinction between a continuous and a discontinuous mode of evolution is by no means negligible. The standard synthetic theory knows no sudden instabilities and sudden phase changes; the theory of punctuated equilibria, however, affirms them. The new theory recognizes the occurrence of long periods of stasis, during which the catalytic cycles that maintain organic species in their environments perform adequately and correct for a limited range of perturbations, and it claims that when the epochs of stasis come to an end, evolution is sudden and unpredictable in detail.

According to this "saltatory" theory, the rate of speciation is strongly influenced by the survival patterns of competing species. Species capable of surviving under a relatively wide range of environmental conditions (including a variety of climates, topologies, predator and prey populations, etc.), speciate less often than those which are fitted into narrower niche structures, or "environmental ruts." When many highly specialized species coexist within a clade in ruts that differ only slightly from one another, relatively small changes in the niche structure can knock out particular populations. "Specialists" speciate under changed conditions where "generalists" can still adapt and survive. The diagram of the branching tree of life no longer resembles the continuous Y-shaped joints of the synthetic theory; it is now pictured in terms of abrupt switches, from dominant species which become extinct to hitherto peripheral ones that become dominant, with short life lines for specialists and comparatively long ones for generalists.[2]

86 THE REALMS OF EVOLUTION

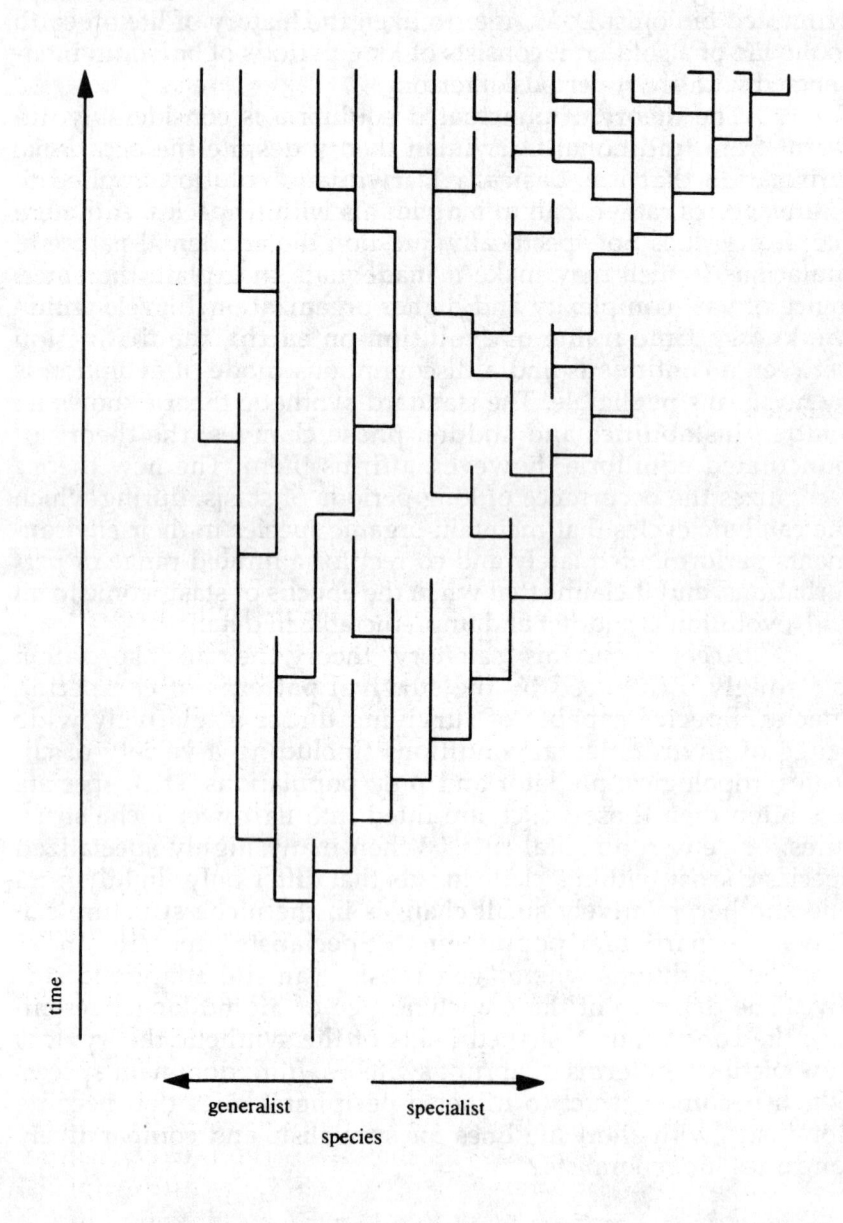

Figure 4.1. *Speciation in biological evolution*

Not only do species appear in sudden transformations within a clade, but entire genera make their appearance in bursts of creativity that mark given paleological epochs. According to recent conceptions an explosion was triggered when the previously dominant community of algae (consisting of prokaryotic cells) was destabilized by the appearance of single-celled eukaryotes that fed on the algal community. By cropping the algae they broke through the epoch of stasis that had persisted for billions of years. The algae were destabilized; niches for additional species were created; and subspecies that emerged on the periphery could move in to occupy them. A large variety of prokaryotes appeared and these in turn made possible the emergence of more specialized eukaryotes; these functioned as their "predators."

Mutation

In order for new species to emerge not only need new niches be available, also mutant populations must be at hand to exploit the available niches. Mutation, an essential element in the Darwinian synthetic theory, remains a cornerstone of the new biology. While proponents of the theory of punctuated equilibria no longer view the emergence of new species as a direct consequence of natural selection acting on mutant individuals, mutations transmitted by reproducing individuals remain a key factor in creating the alternative species or subspecies that achieve dominance when the previously dominant species in a clade become critically destabilized.

Contemporary microbiology finds that mutation is a considerably more complex process than previously envisaged. The process is open to influence from the environment, and it is saltatory, occurring through a phase change in the genome.

The genome—the specific genotype of an individual—contains the genetic program with the necessary information to interact with environmental factors within and external to the organism and to determine the (permanent or transient) characteristics of the phenotype. Biological organisms are maintained in the environment through the reliability of the information coded in the genome. The phenotype (the actual organism) is the outcome of the interaction between the information-based instructions carried by the genotype and the relevant features of the external environment. A mutation, then, is an alteration of the individual's genome that in turn produces an altered phenotype and thus alters the genotype of the species.[3]

While each species has the capacity to transmit the content of its genotype from generation to generation, assuring the reproductive invariance of its species, mutations can change the sequence of nucleotides in the genome of any given individual. The nucleotide strand, arranged as a double helix, is endowed with considerable degrees of conformational flexibility. The folding of the nucleotides has to be extremely accurate; for example, in the average human diploid nucleus the length of the nucleotide sequence is about 1.7 meters—and this sequence is fitted inside a nucleus with a radius of 2.6 micrometers! (The length of the strand in the double helix is 650,000 times the radius.) Recent findings indicate that the required accuracy is not rigid: different configurations are possible and physical factors, such as humidity, temperature, and ionic strength, among others, exercise an influence.

In higher organisms only about 20 percent of the nucleotide sequence of the double helix is sufficient for assuring the coding and regulatory function of the genome. The remaining 80 percent appears to be useless for such purposes; in fact it is often called the "silent DNA." The silent segment consists of a high degree of repetition of certain sequences with a scattered distribution throughout the genome. This may be crucial in mutations since it may allow sudden genetic variation through a translocation of elements in the nucleotide sequence and through quantitative changes by means of an increase or decrease of nucleotides. The genome, it appears, mutates through sudden phase changes induced by changes in the silent DNA.

The nature of transformations in the genome is the subject of intense investigation. Some scientists envisage shocks faced by the genome to which it responds in a programmed manner. Sensing mechanisms, it is believed, alert the genome of imminent threat, and the genome responds by restructuring itself, assuring its own survival as well as that of the cell. Such restructurings produce mutant individuals which in turn can lead to speciation when populations of mutants enter into competition with other species within a clade.

The genome itself proves to be a nonlinear chemical system far from equilibrium. It maintains its structure in a continuous flow of matter and energy through numerous self-regulatory and replicative processes, including autocatalytic cycles. And it undergoes sudden phase changes when challenged by "shocks" or other destabilizing changes in the values of some critical parameters.

The new findings of microbiologists concerning the nature of mutations, and the new theories of paleobiologists concerning the tempo and mode of evolution, produce fresh insights into the dynamics of biological evolution. There is still an element of chance, but there is also pattern and regularity. The response of the genome to environmental perturbations cannot be predicted in detail; likewise the outcome of the macroevolutionary processes of speciation remains generally unpredictable. However, randomness is not unmitigated. The laws that govern biological evolution conserve their validity throughout the range of individually unpredictable processes. Biological evolution unfolds in a lawlike manner in that, independently of specific detail, when changes in the niche structure critically destabilize a dominant species in a clade, the destabilized species yields dominance to the mutant species or subspecies that have been produced by prior phase changes in the genome.

Convergence

In order to understand how evolution climbs ever higher on the cluster of organizational levels in the realm of life, we have to add another factor to speciation and mutation: the factor of convergence. Convergence—the tendency of third-state systems to form hypercycles in a shared milieu—introduces the element missing in mainstream biological theory and removes the puzzle of how high levels of complexity can be achieved in relatively short time frames. Hierarchization through hypercycles—in other words, the principle of convergence—accounts for biological evolution's climb to higher organizational levels: it shows that the outcome of the processes of mutation ant speciation is not entirely random but follows a pattern observed in all the realms of evolution.

The effects of convergence can be traced throughout biological evolution. Already in the early history of life, eukaryotic (nucleated) cells evolved through convergence among prokaryotic (non-nucleated) cells. Many of the vital organs of eukaryotes, such as mitochondria, chloroplasts, flagella and the mitotic apparatus, may have originated as hypercycles formed among primitive prokaryotes. Multicellular species, in turn, emerged out of hypercycles among the eukaryotes. On still higher levels, entire ecosystems resulted from hypercycles among muticellular species. Life is continually exploring novel combinations of structures and functions as existing species interlock their catalytic cycles in shared habitats and jointly converge in higher-level systems.

The benefit of evolutionary convergence toward high organizational levels is the dynamism and autonomy available to multilevel systems. Roughly at the level of protocells, autopoietic systems appear—the hypercycles evolve the ability to replicate the entire cell structure. Cell division renders single-celled organisms theoretically immortal: they can replicate themselves as long as boundary conditions remain favorable. Additional autopoietic functions appear when species converge on the multicellular level of organization. The reproduction of the entire organism has been "invented" to compensate complex nonequilibrium phenotypes for their individual mortality. Reproduction in the bisexual mode has become a source of variety within species; through the mechanism of recombination it assures that there are individuals fitted to a large variety of niches.

In the trend toward dynamism and autonomy the key innovation, for which the multicellular level of organization is sine qua non, is endothermy: a constant and high internal body temperature. The typical temperature of 37° C stores considerable free energy and allows a wide range of activity with speed, variety, and stamina. A constant input of free energy-containing substances and a dependable elimination of waste products maintains the homeostatically stabilized "milieu interieur" of complex warm-blooded organisms. The entire metabolic process is precisely regulated on the highest, controlling level by catalytic cycles and norm-maintaining negative feedbacks, enabling the organism to compensate for variations in its external environment.

While the benefit of evolutionary convergence is increased dynamism and autonomy, the cost is increasing vulnerability. For example, the body temperature of mammals is close to the upper threshold of permissible temperatures for most living protoplasm. Also, the survival range becomes dramatically reduced: while a cold-blooded organism can tolerate a wide range of temperature conditions (some fish can recover even after being frozen), warm-blooded animals must be able to maintain their body temperature within a few degrees from the norm, or succumb.

Convergent evolution to higher and higher organizational levels involves a gamble: the exchange of relatively simple and reliable catalytic cycles for complex, hierarchically organized sequences of dynamic individual- and species-maintaining hypercycles. The kind of stability typical of the simplest forms of life (such as blue green algae and other prokaryotes) is surrendered, and its absence is compensated by highly sophisticated feedbacks coding multilevel self-regulating mechanisms. The gamble means greater dynamism and autonomy—but at the cost of the mortality of the individual and the risk of sudden destabilization and ultimate extinction of the species. The fossil record testifies to the poor odds in this gamble: more than 96 percent of the biological species that at one time populated this planet have ultimately disappeared.

The arrow of time in biological evolution points toward increasingly dynamic and autonomous species. At the same time it points toward species that are highly complex and vulnerable, forced to rely on delicate survival functions that include, in higher species, not merely genetically coded and inherited behavior, but also learning acquired in the lifetime of the individual. The gamble of advanced evolution can pay off—for a time at least. *Homo* is living proof of it.

THE EMERGENCE OF *SAPIENS*

Although there is no single universally accepted theory of the origins of our species, it is probable that the ancestral tree of the family of man exhibits the same punctuated, saltatory mode of evolution as other forms of life. The divergence of hominids from the apes sometime between four and eight million years ago is itself a relatively sudden event: within a matter of a few million years an erect, bipedal species emerged, leaving behind the tree-dwelling apes.

The mutation may have been due to increased dryness during the Pliocene when East African tropical forests receded and were replaced by more open savannahs. An upright posture would have been advantageous in tall grasses, and bipedalism would have freed the arms for the use of tools and for wielding weapons in confrontation with large, savannah-dwelling predators. The earliest hominids may have spent some of the time in the trees and some on the ground, moving toward full bipedalism in stages. In any case the ancestors of *Homo* were full-time bipeds already two million years ago.

Our direct ancestors are likely to have evolved through a rapid and saltatory speciation—a true bifurcation—within the hominid group. There is no firm evidence of gradual evolution within any hominid taxon. *Sapiens* does not appear to be the result of a gradual series of adaptive mutations within a single lineage such as *africanus—robustus—boisei—sapiens*. It seems likely that neither *robustus* nor *africanus* evolved into *boisei*, nor indeed did *boisei* evolve into either *africanus* or *robustus*. Paleoanthropologists are faced with a number of puzzles in the interpretation of recently discovered fossils, but seem generally disposed to regard *afarensis* as the common ancestor of *sapiens* and other, now extinct hominid species. Should this be the case, some three million years ago *afarensis* would have speciated into four distinct hominid lineages (*habilis—erectus—sapiens*, *robustus*, *africanus*, and *boisei*) and have done so within the remarkable brief span of 300,000 to 500,000 years. For now scientists can only speculate on the reasons for this rapid radiation: it could have been caused by a change in the climate, by changes in competing species, or some other variation in essential parameters. In any case, *afarensis*, the then-dominant hominid species, seems to have been rapidly destabilized in its environment and replaced by various mutants that emerged on the periphery. If the hypothesis is correct, then the lineage that gave rise to *Homo* exhibits the same kind of saltatory bifurcation that is characteristic of the evolution of all systems in the third state, whether they are in the physical, the biological, or—as we shall see—even the social realm.

Once a species has burst on the scene, it appears to have remained basically unchanged throughout its existence. *Erectus*, our direct ancestor, lived in various geographic locations with but minor variations for a million and a half years. In its much shorter span, *sapiens*, too, has been essentially unchanged. The seeming continuity of evolution hides a series of discontinuities, as species disappear and others take their place. In the case of the hominids, it was *sapiens* that survived, while the others became extinct.

We now have reasons to believe that modern man became a dominant species only recently. While our ancestors were present in Africa already 100,000 to 130,000 years ago, it was only some 30,000 years ago that they achieved dominance in Europe. It was around then that Neanderthal man disappeared, and his disappearance, while still shrouded in mystery, may have had much to do with the appearance of sapiens. There is no record of any hominid species but *Homo Sapiens* in Europe during the last 30,000 years. The fossil record indicates that sapients migrated from Africa about five thousand years beforehand and may have initially coexisted with the Neanderthals, possibly on different locations. But as sapients spread into East Asia and as far as Australia, the major competing species vanished, creating another puzzle for anthropologists. Did the sudden shift in Europe come about because sapients could better adapt than Neanderthals to the warming trend that made for the appearance of forests in place of the great steppe? The answer is not clear. Yet the fact seems indisputable: sapients invaded from Africa and became the sole hominid species in Europe within the brief span of five thousand years. This is yet another instance of saltatory evolution through the destabilization of a dominant population and the subsequent rise to dominance of a species invading from the periphery.

Changing conditions in the biosphere favored the preservation of the traits of *Homo sapiens* and his spread as a dominant species to the inhabitable parts of the continents. A species capable of conscious thinking, symbolic communication, and complex social organization emerged as the supreme predator of life on earth.

But the appearance of modern man was not without risks: our species entered into the evolutionary gamble in a grand way. It entrusted its individual and species survival more and more to the perception and interpretation of the environment, and less and less to intrinsically stable physiological mechanisms and genetically coded behavioral patterns. So far, the gamble paid off: *Homo* is still alive and dominant. But he now lives within sociocultural systems that he created but no longer knows how to control. His future will be decided by the evolution of these still higher-level systems—more exactly, by his ability to evolve the power of his brain and mind to steer the course of their evolution.

Although biological evolution on earth culminated in some sense in *Homo*, there is nothing predetermined about the appearance of our species—no final cause or general plan that would lead to man by design or by necessity. Biological evolution, like the evolution of all third-state systems in the empirical world, does not determine the course of its unfolding but merely identifies the possibilities and sets the constraints. Notwithstanding the many specific features of the evolution of life, it shares its basic mechanism with all other forms of evolution. Speciation, the sundering of one coherent, integrated and reproductive community into two (or more) species, is a bifurcation in the evolutionary path of a lineage. Similarly to bifurcations in all dynamic systems, the outcome of individual speciation processes are indeterminate and unpredictable; all we can tell is that the emerging species will be better fitted into its milieu (that is, will have greater reproductive success) than its predecessor. But despite the individual unpredictability of the particulars of emerging species, there is an overall, statistical predictability associated with the historical course of macroevolution itself: the set of all speciations in the last four billion years manifests a definite convergence toward higher organizational levels. This process does not necessarily lead from protocells and algae specifically to *Homo*, but it does lead from systems that are relatively simple, microscopic, and comparatively close to equilibrium toward systems that are further from equilibrium, larger in size, greater in complexity, and more dynamic and autonomous.

Evolution is not teleological; it does not have a precise goal in the form of any particular species of organism or ecosystem. Yet it is directional in that it drives systems in the third state progressively further from equilibrium into the high-level and nonlinear realm where life appears and then mind and intelligence.

Notes

1. In the classical account, natural selection acts on essentially random mutations. Mutations are "typing mistakes" in the repetition of the genetic code of the parent in the offspring, produced by all species at a more or less constant rate. Most of the mutants produced by chance variations are faulty in some respect and are eliminated by natural selection. However, random mutations occasionally hit upon a genetic combination that renders the individual more rather than less fit to live and reproduce. It transmits its mutant genes to successive generations, and the comparatively numerous offspring produced by these generations deplace the previously dominant species. The range of possible variations is constrained only by the comparative fitness of the mutants for life and reproduction in their habitats.

2. For example, a generalist species such as the *Aepycerotini* (of which the living member is the impala) seems to have survived the last six million years with only one or two speciations. But the more specialized *Alcelaphini* (which includes hartebeests and wildebeests) has undergone twenty-seven speciations during the same period: changing system-environment relations kept knocking out the specialized populations from their narrow environmental ruts and replacing them with others more fitted to their new niche structures.

3. Genomes, consisting of nucleic acids (mainly DNA and RNA), perform numerous functions including self-replication, recombination, self-regulation and self-repair, and the synthesis of messenger RNA for the formation of proteins. All functions are coded by sequences of four nucleotides (adenine, guanine, cytosine, and thymine for DNA, and uracil instead of thymine for RNA) folded as a double helix inside the nucleus.

5

The Evolution of Society

The hypothesis that the laws governing the evolution of natural systems also govern the development of human societies is contrary to a humanistic tradition in the social sciences, but it is intrinsically reasonable. These laws do not prescribe the course of evolutionary development but merely set the rules of the game—the limits and the possibilities that the players themselves exploit. The ground rules for biological evolution have been set by evolution in the cosmos. For sociocultural evolution the rules have been set by biological evolution, first and foremost by the evolution of the brain and mind of *Homo sapiens*.

       The evolutionary thesis does not mean that human societies are biologically determined. It means only that societies are evolving systems emerging and persisting within the multiple-level structure of other systems in the biosphere. Societies follow the rules set by the general laws that govern the evolution of such systems within the limits and possibilities created by human beings, their values, beliefs, habits, and mores. But societies follow these rules on their own, typically societal level, and not on the biological level of their members.

The evolutionary thesis, though not biologically reductionist, does draw on concepts and theories developed in part in the natural sciences. It thus raises the specter of perennial controversies between natural and social scientists concerning the application of the postulates of the natural sciences to the sphere of human affairs. Historians, especially, have traditionally considered their field a record of historical events and not the exemplification of universal processes. History, they say, deals with the unique, the individual, and the concrete; it also deals with events infused with human will and purpose. By contrast, the natural sciences deal with the universal, the general, and the abstract, and their subject matter is untouched by conscious purposes. The natural sciences postulate laws to predict and if possible control phenomena; but prediction is impossible in the realm of history and control is often undesirable.

As already noted in Chapter 1, in the history of Western civilization the split between the natural sciences and the humanities goes back at least two hundred years. But this split is now transcended by fresh knowledge coming from the sciences of complexity, laying the foundations for a unified, grand conception of the evolution of third-state systems in physical nature, in living nature, and in history. Regrettably, humanistic historians and social scientists are seldom up to date on these developments. The scientific laws known to the great majority of historians are deterministic and mechanistic; they are basically the laws of Newtonian physics. Historians are justified in objecting to their application to the events and processes of history. However, the laws of evolution are not mechanistic and deterministic, and thus —in principle at least—they can be applicable to human beings and to their social structures and processes.

In any case, the applicability of evolutionary concepts and laws to the development of human society should not be prejudged. Whether or not humanity changes and develops according to the laws that govern the evolution of natural systems cannot be answered by abstract logical analysis. In order to decide the validity of the evolutionary thesis we must confront the basic concepts of the sciences of complexity with historical facts and empirical observations. While this project obviously cannot be completed on these pages, we can outline the basic evolutionary axioms of a theory of society and explore whether or not they apply to characteristic and significant features of historical development.

THE EVOLUTIONARY AXIOMS OF SOCIETY

1. Society is a system composed of groups of human beings in specific relations. The social system—much like an organic population, a clade, or an ecosystem—maintains itself or changes independently of the particular destinies of its individual members. Human beings pass through it in cycles of birth, maturation, and death; society persists, develops, or decays according to processes that take place on its own societal level. The fact that the individual members of human societies are conscious, language- and tool-using persons specifies but does not determine its evolutionary dynamics; it only introduces the specific control parameters that set the range of possibilities and the nature of the constraints in the behavior of the social system.

2. A dynamic system in the third state, society occupies a cluster of organizational levels above those of third-state systems in the physical and biological realms. The social system's environment is both natural and social: it is in part the biosphere and its various ecologies (with the sun as the main energy source and surrounding space as the main energy sink) and in part the sociosphere, made up of other societies and their infrastructures.

3. Society is neither a natural system, like an atom, a molecule, or a cell, nor an artificial system such as a machine or a computer. It is the result of human action and interaction, but not of conscious human design. Unlike specific organizations (corporations, public and private institutions, armies, churches, professional associations, social clubs, and so on), the networks of relations that bind individuals in entire nations, states, and cultures cannot be designed; they emerge in the course of history. The degree of complexity attained in a modern society, though modest in comparison with an organism, exceeds by far that which its human members could achieve by purposeful design.

4. Society, though composed of human beings, is not reducible to the sum of their behaviors and attributes; it evolves functions and acquires attributes that are typical of its own, societal level of organization. Such spontaneously emerging features include the capacity to replicate the principal components (public institutions, enterprises, and other groups and associations of individuals) and to renew any part of the structure that may have been impaired by internal or external fluctuations. Self-organization is also a fea-

ture of society. Like other third-state systems, society is a self-evolving system in its own right, capable of settling into alternative steady states following critical perturbations. Through autopoiesis and bifurcations, society maintains itself in its particular milieu and, if viable, evolves new structures and organizational forms in the course of time.

5. Notwithstanding the suprabiological organizational levels of society, its structural complexity—though far greater than that of artifacts—is less than that of its individual members: the human brain alone is more complex by several magnitudes than all the societies of the world put together. The relative simplicity of sociocultural systems is consistent with general evolutionary trends. Systems on a higher level of organization are initially always simpler than the systems that constitute their main components; a new organizational level creates a simplification and not a complexification of system function. (The comparative simplicity of sociocultural systems is readily explained by comparing the biological time scale with the time span of social history. The origins of the genus *Homo* go back millions of years; the origins of human societies as sociocultural systems, rather than as tribal reproductive units, go back merely twenty or thirty thousand.)

6. Society evolves through convergence to progressively higher organizational levels. As the flows of people, information, energy, and goods intensify, they transcend the formal boundaries of the social system. The catalytic cycles that maintain the system in its environment encounter similar cycles in the intersocietal milieu and interact with them. In time the cycles achieve coordination as intersocietal hypercycles. Thus neighboring tribes and villages converge into ethnic communities or integrated states, these in turn become the colonies, departments, provinces, states, cantons, or regions of larger empires and ultimately of nation-states. When empires are stripped of their far-flung territories and overseas colonies, the capital regions and the liberated states are open to new forms of convergence among themselves. In today's world this process leads to the creation of various regional and functional economic and political communities and blocs among developed as well as developing nation-states.

The evolutionary axioms apply to society because society forms a third-state system on the suprabiological cluster of organizational levels. Society is such a system since human beings have evolved as social creatures: they tend to behave with some measure of coherence. Coherence results not because of some metaphysical property that makes humans into collaborative social animals, but because prolonged interaction among individuals leads to the emergence of an overall order that is independent of the conscious aim of any of them; and this order is conserved and handed down to successive generations through the patterns of value, belief, and custom prevalent in society.

The historically evolved social orders arc constantly shaped by individual action and interaction and modified by changes in collective culture and public policy. The orders impose constraints on individual behavior, although these constraints are perceived as such only if they are out of phase with the values, expectations, and general cognitive map of individuals. In a relatively stable, unstressed state of society, the constraints imposed by the spontaneously evolved social order appear as accepted forms of social intercourse. They bond individuals within kinship, community, or interest groups. In traditional societies myths and religion create the main kinds of social bonds, while modern societies produce many types of bonds and allegiances, cemented not only by mores and customs but also by legal and juridical systems and rules governing public and personal behavior.[1]

THE DIRECTION OF SOCIAL EVOLUTION

Society, a dynamic but relatively simple suprabiological system, evolves in a fuzzy and generally disordered manner—so much so that students of history often question whether there is any pattern underlying the sequence of historical facts and events. History, as positivists claim, seems to be "one darn thing after another." Yet if we place the here outlined axioms in the context of a sufficiently large cross-section of historical development, not only can we perceive basic patterns in history, but we can perceive that these patterns arc consistent with the general direction and the dominant dynamics of the evolution of life in the biosphere—and likewise with the evolution of matter in the cosmos.

The people of ancient civilizations have perceived patterns in the way societies exist and evolve millennia before social scientists ever posed the question. Agrarian societies. strongly dependent on the weather and thoroughly integrated in their natural environment, believed in a cyclical pattern, following the seemingly eternal recurrence of the seasons. Classical societies sometimes perceived themselves as having descended from an earlier golden age: the pattern they perceived was no longer circular but traced a descending slope. When notions of progress became more firmly established, societies qualified the eternal recurrence of the circular pattern with the image of an ascending helix: all things recur, but they do so on a higher level. It was only in modern times that the dominant pattern in historical development was conceived as an arrow with an irreversible, upward direction of flight.

The pattern indicated by the evolutionary hypothesis is progressive but not linear. It resembles a fluctuating graph with many local peaks and valleys but with an ascending tendency. The state from which, and that toward which, it ascends can be best elucidated by comparing the earliest known stage of historical development with the latest stage. The progression from Paleolithic to modern times is highly uneven, with countless forward leaps and sudden regressions, yet it exhibits an overall direction. The structure of Stone Age tribes is considerably less complex than the structure of modern nation-states. More important, as we shall see, Stone Age tribes are systems with a far more limited capacity to access, store, and use dense fluxes of free energy in their environment than modern societies.

It is clearly an oversimplification to base arguments for directionality in history on a comparison between Stone Age tribes and modern nation-states. But these two types of social systems represent the temporal poles of societal development known in history. Recorded history begins with Stone Age tribes and extends, to date, to modern nation-states. In the intervening epochs, despite temporary regressions and deadends, structural complexities have grown and the dynamism and autonomy of societies have increased.

Technology as Engine of Change

The question of the motive force behind the observed general progression in history can be answered relatively unambiguously. Although the progression is not smooth and continuous, and it is not predetermined in any of its phases, it does exist, and its engine is technology. Technology has to be broadly conceived—as is customary today—as the instrumentality that imbues all human activities and extends human powers to act on nature and interact with others.[2] A technological innovation is not just the invention of a tool, but the stretching of the imagination and the transformation of common sense. A major technological breakthrough makes the supernatural natural—as, for example, with the mastery of fire and then of flight—and renders the abnormal and the unthinkable normal and even commonplace—as with a nuclear reactor or the instantaneous transmission of image and sound. It challenges people's values and practices and shakes the foundations of established institutions.

The impact of technology on society is generally proportionate to the flexibility of the dominant modes of social organization, that is, to the ability of society to apply internally or externally generated innovations. Stone Age societies were enduring but extremely rigid; they had little capacity for adopting (and perhaps even for coming up with) innovation of any sort. They maintained themselves by a fixed system of rites, rituals, and taboos, and corresponding myths and belief systems. The Paleolithic or Old Stone Age takes up much of the history of the hominid species, beginning first in Africa during the Lower Pleistocene some 1.75 to 2 million years ago. It took perhaps 1 million years for its primitive toolmaking to spread to Europe and Asia. The earliest technologies were simple physical implements: natural stones with little or no modification, or stones unifacially or bifacially flaked for chopping, bashing, or pounding, with some flakes for skinning and scraping. The social structures were limited to small family groups hunting and gathering together in open savannah country.

By the Middle Pleistocene human artifacts became somewhat more sophisticated and specialized: there were cleavers, points, hand axes, choppers, scrapers of various sizes, points, and related implements. In the early Upper Pleistocene hunting became selective, and more functional implements appeared: wooden spears and throwing and digging sticks, and in some regions the use of animal bones and antlers as tools. But it was only during the last 35,000 to 40,000 years in the Upper Paleolithic that greater advances were made, including knife blades, projectile points of stone and bone, antler for spears and darts, needles, chisels, spearthrowers, and harpoons, among others. The bow and arrow came into use toward the end of the Paleolithic as hunting became specialized and geared to the seasons and one or two species of prey.

Through nearly two million years, the precursor species of *Homo sapiens* lived a stable existence in close symbiosis with the environment. Except for causing the extinction of a few intensely and efficiently hunted species toward the end of this period, the ancestors of man did not ruin or unbalance either their environment or their own social structures and practices. The size of the communities remained small, and the division of labor may have been indistinct, as hunting and gathering are likely to have been cooperative activities that left little time for more specialized pursuits. But the slow yet persistent improvement of Stone Age technologies permitted populations to spread over vast, in part previously uninhabited areas.

The size of communities and their ways of life changed markedly with the advent of the Neolithic, around 8,000 years ago in the Levant. The cultivation of grain and the domestication of sheep and goats heralded a major technology-triggered revolution in society. It did away with the need to follow the source of food in small bands and transformed social rituals, values, and beliefs. Nomadic tribes were settled in permanent abodes and could grow more populous; a growing segment of the population was freed from the arduous tasks of hunting and gathering food and could take up social and even cultural pursuits. By the seventh millennium B.C. these advances had spread into eastern Europe, and by the third millennium up the Danube and across to the Atlantic coast. While the new technologies appear to have spread into the Nile Valley and India as early as the fifth millennium B.C., they had no impact on the lifeways of large segments of the populations of eastern Asia and sub-Saharan Africa until about the first millennium B.C., owing to the bountiful environment provided by savannahs and forests.

A technological breakthrough with equally vast consequences proved to be the invention of writing. It made possible the transmission of knowledge and experience, and the creation of far-flung empires with complex systems of taxation and sophisticated administrative structures. It also accelerated the spread of the other technologies, and the institutions, values, and beliefs that accompanied them.

In the span of the last millennium the tempo of social change has greatly accelerated. The effects of technological innovations have been amplified by rapid means of transportation and communication and exported by dominant economic powers and conquering armies. The archaic empires could maintain themselves for centuries—some for millennia—in favorable natural environments, mainly in river valleys such as the Nile, the Ganges, the Tigris-Euphrates, and the Huang Ho. But empires invented not only the means to produce food and administer far-flung territories; they also invented the technologies of conquest and subjugation, supplementing the power of human muscle with lances, arrows, and swords. Empires spread their dominance over ever-vaster distances and greater populations.

As William McNeill points out in his classic *The Rise of the West* (1963), the early structures of civilization displayed essentially the same characteristics wherever they evolved. They were agriculturally based hierarchies organized around cities and united into empires. They made use of written languages, had religions that were employed even more for political than for spiritual purposes, and maintained armies as well as modest scientific and technological establishments. The four or five empires that dominated civilization in the ancient world retained these basic structures for millennia.

In the Middle Ages, however, the classical structures broke down at the western edge of the civilizational pattern. In Europe, new technological developments, the relative weakness of the form in which the pattern was embodied, the peculiar attributes of the barbarian invaders, and certain features of the Greco-Roman tradition conspired to destabilize societies. During the period that encompassed the Renaissance, the Reformation, and the emancipation of science, European societies underwent a major transformation.

The causes of instability in Europe were multiple, some with long historical roots. For a long time after the ending of the barbarian invasions, Europe had enjoyed a modest but unrelenting growth in population. By the middle of the thirteenth century the bonds created under feudal society became strained. When the Black Death struck in 1348, a period of economic expansion was suddenly arrested and many of the bonds were permanently ruptured. The destabilization of European society is recorded in the attack of nominalist philosophers on the hitherto uncontested medieval world picture, and in the progressive independence from papal power of the political institutions nourished by the church for its own purposes. As technological innovations such as the compass, sturdy sailing ships and gunpowder were introduced, voyages of discovery extended the spiritual horizons of medieval Europe. They soon gave way to voyages of conquest, which in turn led to economic and political expansion.

Europe in transformation became the force that propelled the rest of the world out of its classical stability and into a new and uncertain future. Europe itself was in turmoil; the way to the modern age was not found without chaos and experiment. Religious reformers attempted to reestablish a pristine form of

Christianity, while Renaissance scholars strove to revitalize civic humanism in the classical mold. But there were also scientific minds such as Galileo, Bruno, Copernicus, Kepler, and then Newton, who spearheaded a movement that led to the emancipation of science from medieval dogma. Modern science, with its insistence on observation and experiment, created a "purified" image of the world that, while perhaps simplistic in its mechanistic assumptions, proved to be eminently suitable for practical applications. As European societies realized the potential of autonomous science to "wrench the secrets of nature from her womb" and transform nature for human ends, a new technology-impelled revolution got under way.

The first industrial revolution created new relations between man and man and between man and nature, destabilized established institutions, and reformed traditional value and belief systems. It displaced traditional occupations and opened up new ones. The villages lost ambitious young people and impoverished peasants to new urban and industrial centers, where the former could occasionally rise on the ladder of status and wealth while the latter were forced into the sweatshops and "dark satanic mills" of the new industrialists. Gradually, through slow and often painful adjustments, better institutions emerged. They restabilized medieval princedoms, kingdoms, and tsardoms as modern industrial societies.

Hardly consolidated, the industrial revolution was rocked by two world wars. These cataclysmic events, in addition to loosening established orders and institutions, promoted the discovery and application of several new technologies that soon moved from the military to the civilian sphere. The new technologies of information, automation, and communication heralded another transformation: the second industrial revolution. Its impact on society was as profound as that of the Neolithic revolution, and infinitely faster. Complexity, based on the availability and rapid communication of information, grew in every domain. A tertiary sector of services emerged beside the traditional sectors of agriculture and industry. The mass media achieved an unprecedented level of influence. Wherever they penetrated, the technologies of information, automation, and communication revolutionized every aspect of life in society

In the span encompassed by Paleolithic societies on the one end and modern information-based societies on the other, an entire succession of societal forms has unfolded. The nomadic tribes of the Paleolithic transformed into the settled villages of the

Neolithic; these in turn gave way to archaic empires and to local kingdoms and city-states. The classical empires were followed by medieval princedoms, and these yielded to the rise of nation- states, some with vast colonies. Today the colonies have disappeared, and modern nation-states have spread to the four corners of the world.

This historical succession of societal forms and types is capable of myriad interpretations, depending on the features and factors one chooses as one's point of departure. In light of the evolutionary axioms, the basic factor is not the variation in the adopted political, legal, and economic systems. These systems may vary in any historical epoch between the opposite poles of distributed power and democracy on the one hand and concentrated power and dictatorship on the other, between individual ownership and unconstrained markets on the one extreme and heavily structured and constrained forums of exchange on the opposite. Rather, the basic and noteworthy factors in the historical progression are the ways a society performs its autopoietic and evolutionary functions. These factors concern the interplay between technology—broadly defined—and the established social institutions and practices. The historical succession is best conceptualized in terms of the transformation of dominant technological types, and the structural and institutional changes catalyzed by them.

With attention to both the technological and the social factors we can perceive a series of dynamic transformations in the development of societies. Nomadic hunting-gathering tribes domesticate plants and animals and transform into settled agrarian-pastoral societies; agrarian-pastoral societies evolve such technologies as irrigation and crop rotation and transform into agricultural ones; agricultural societies develop handicrafts and simple manufacturing technologies and thus transform into industrial societies; and industrial societies, under the impact of new, mainly information and communication-oriented technologies, evolve into postindustrial societies.

The evolutionary vision perceives history's arrow of time flying along the axis: hunting-gathering → agrarian-pastoral → agricultural → preindustrial → industrial → postindustrial society. The flight of the arrow may be interrupted at any point; it may be temporarily halted at any point; and it may be made to skip a stage at any point. But it may not, except under the impact of unusually powerful external factors—enduring foreign domination, long-term climactic changes with adverse effects, and the like—be fully and steadily reversed.

There is, in consequence, a sense to the terms *progress* and *development* in history. They denote an ineluctable advance of society along an evolutionary axis, sparked by technology. Technological change pushes societies along the here-noted trajectory, for technological change is irreversible on the whole. Societies progress mainly by creating, assimilating, or adapting technological revolutions; and technological revolutions, whatever their precise nature, always move from the hoe to the plow and not the other way. Only those technological inventions are adopted and handed down which represent an improvement in the efficiency of some procedure: greater speed, less investment of time and money, and operation on a larger scale. This was true throughout the range of technological innovation from the wheel of agrarian society to the steam engine of industrial society, and it remains true today in the passage from the diesel engine of industrial society to the silicon chip of the postindustrial one.

However, the adoption of a new technology did not and does not necessarily mean an immediate advance in energy intensivity and efficiency. Classical Greece and Rome, for example, knew many more technologies than they actually employed; the institution of slavery discouraged the introduction of energy- and labor-saving devices. Although there were mines, clocks, and gadgets of various kinds, neither animal power nor the power of water and wind was effectively harnessed. Not even the vacuum and water pumps devised by Hero were used to alleviate human labor. On occasion specific pecuniary or political considerations blocked the adoption of a new technology. Emperor Tiberius destroyed the formula for glass that was said to be unbreakable and had its inventor put to death; he feared that unbreakable glass would undermine the value of gold, silver, and precious stones. Emperor Vespasian paid the inventor of an efficient system of pulleys and levels designed to lift and carry stone pillars but then had the model wrecked and prevented its use: he was concerned that the labor displaced by the device would create social unrest.

Economic, social, and political considerations can prove to be stronger than considerations of efficiency in use: also, the dominant values of a civilization can discourage the practical application of technological inventions. Gunpowder was invented by the Chinese but was used only in fireworks to mark ceremonial occasions; and the compass, another Chinese invention, had to wait for the coming of Western navigators to find practical employment. Neither the classical Mediterranean nor the ancient

Chinese civilization valued energy efficiency, and both thought the design and use of mechanical devices unfit for soldiers, scholars, and other persons of standing and distinction.

In modern times as well, technological choices are often dictated by considerations other than improvement in energy efficiency. The range of technologies that waste energy and raw material is well known. But it is less widely known that even the gasoline engine, that bulwark of modern life, was chosen over the steam engine for reasons that had nothing to do with its competitive energy efficiency: when the Stanley Steamer and the four-cycle Otto engine competed for acceptance, the Stanley Steamer was unexpectedly handicapped by an ordinance that called for the removal of the water troughs that were to resupply it along public roads. The reason had little to do with comparative efficiency: it was an outbreak of foot-and-mouth disease among U.S. cattle.

Although some technological inventions are never adopted, and those that are adopted are not necessarily the most efficient, all inventions that *are* adopted represent some measure of improvement in an activity or procedure. The refusal of societies to adopt technologies that actually reduce energy efficiency makes for directionality in history. It is the barb that keeps the wheel of societal change from spinning backward; the bowstring that propels the arrow of time forward.

In the late twentieth century we are in an era of growing societal dynamism achieved mainly through gains in energy efficiency. The information and communication technologies of the second industrial revolution do more with less: they allow significant energy savings by making use of far more dense energy fluxes than the technologies of the first industrial revolution. They combine with the exploitation of new and abundant energy sources, such as nuclear and solar energy. A renewable energy-using technological society governed by a computer-elaborated flow of information is both more energy intensive and energy efficient than a society based on the burning of fossil fuels. It is further along the road of evolution than any of its historical predecessors.[3]

The trend toward the exploitation of increasingly dense and abundant energies is likely to continue, despite crises, bottlenecks, and short-sighted policies. Nuclear fusion and solar-voltaic technologies are in development, and they assure access to superabundant sources of energy. Fusion liberates the energy contained in lithium and even common sea water; solar technologies convert the energy of the sun into electricity by the use of an abundant material such as silicon. The conception of the planet as a finite container of free energies is demonstrably false. All matter-energy systems bind energy in some way; it is only a question of human ingenuity to find the way to unlock the systems and make use of the liberated energy.

Because technological innovation in society is on the whole irreversible, the arrow of time in history is consistent with the arrow of time in the physical and in the biological realms of evolution. Technological societies set forth the evolutionary progression toward more dynamic and autonomous systems by exploiting increasingly abundant, increasingly dense energy fluxes through correspondingly more complex social structures. Through alternating epochs of stability, and now exponentially increasing technology-triggered bifurcations, societies in history evolve from the low-energy and structurally relatively simple nomadic tribes of the Paleolithic to the dynamic, autonomous, and relatively complex technological nation-states of the contemporary age—and then beyond, to organizational and institutional forms that we can only dimly perceive today.

THE DYNAMICS OF SOCIAL EVOLUTION

Bifurcations in History

History's arrow of time does not fly smoothly. Although the historical record is always complex and frequently obscure, it gives good reasons to believe that societies, the same as biological species, do not change at all times and in small increments. Rather, the mode of change appears saltatory and intermittent, triggered by external conquests and internal discontent and by technological revolutions that change the pattern of relations between man and man and between man and nature.

The discontinuous, nonlinear mode of historical development has been noted by many historians, including Arnold Toynbee. Indeed, Toynbee's theory of "challenge and response" depicts societal change in essentially evolutionary terms. "During the disintegration of a civilization," Toynbee noted in his 1972 book *A Study of History* (p. 228), "two separate plays with different plots are being performed simultaneously side by side. While an unchanging dominant majority is perpetually rehearsing its

THE EVOLUTION OF SOCIETY 109

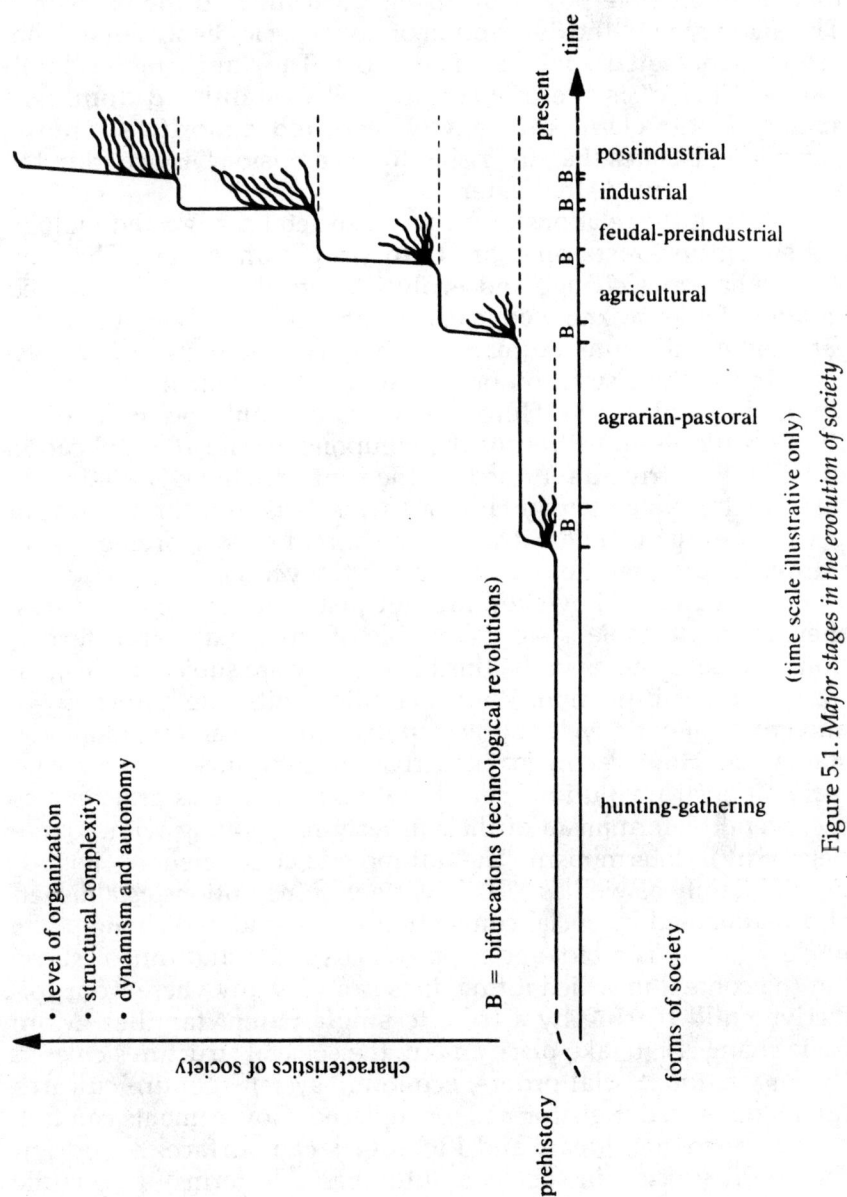

Figure 5.1. *Major stages in the evolution of society*

own defeat, fresh challenges are perpetually evoking fresh creative responses from newly recruited minorities, which proclaim their own creative power by rising, each time, to the occasion." The analogy with the description of how species evolve in the theory of punctuated equilibria is evident. The "unchanging dominant majority" is the unchanging yet destabilized dominant species in the clade; the "newly recruited minorities" are the peripheral isolates that, in "rising to the occasion," move from the periphery to invade the center.

But the relationship between societal change and biological speciation rests on more than simple analogies. The way human societies change and evolve is logical and fully expected in light of current knowledge concerning the evolution of systems far from equilibrium. Human societies, as already noted, are systems in the third state, on organizational levels far from thermodynamic equilibrium. Here structures can only be maintained through the reproduction of the components and the replication of the entire structure formed by their interrelations. Societies are autopoietic systems maintaining themselves in a throughput of people, resources and infrastructures, and a flow of energy, by means of autocatalytic and cross-catalytic cycles.

Societies, however, are not just autopoietic structures; they are multistable systems capable of change and transformation. Like all systems in the third state, they are subject to destabilization and bifurcation. Within certain limits, the autocatalytic and cross-catalytic cycles buffer out disturbances and maintain the status quo. However, major perturbations introduced by wars and technological revolutions can destabilize the cycles and replace the overall determinism of their orderly functioning with a large degree of indeterminism. The autopoietic components— including the family as well as a large variety of institutions established and maintained by social convention—may undergo change. The nuclear family is a biological reproductive unit and thus it stays, but the context in which it functions can vary anywhere from collective child rearing by a tribe to single-parent families. More basic changes can take place among the societal structures created by convention. Social orders, economic systems, entire cultures can be discarded, transformed, or replaced. Governments can fall, new movements, ideas, and ideologies can surface. A sudden change may occur in society's autopoiesis. In terms of dynamic systems theory, society's stable attractors, which hitherto ensured orderly functioning, may turn chaotic or disappear. Society may enter a phase of subtle or catastrophic bifurcation.

A human society is a dynamic entity; even when it is stable, it fluctuates around certain norms determined by the laws, conventions, and behaviors of its members and by their relations with other societies and the environment. Like other complex dynamic systems, at best society can control its own fluctuations and maintain itself in a dynamic steady state. Stability in a society means that autopoiesis is operative: the system can ensure the reproduction and replenishment of its subsystems and its flows, including natural resources, raw materials, foodstuffs, energies, money, information—and people. Bifurcations occur when the steady state can no longer be maintained, and autopoiesis is replaced by a condition of critical instability. In this condition some or all of the flows threaten to be depleted or cut off, and all or some of the subsystems are doomed to atrophy. An unstable society cannot ensure the replenishment of its natural resources, its food and energy supplies, and its wealth, nor can it ensure the survival and reproduction of its people.

The difference between stable and unstable societies is spelled by the way the economic, social, and political systems of production, consumption, and governance operate. These societal subsystems are operated by human beings, but they are not under the full and conscious control of any individual. Rather, the production, consumption, and governance systems function according to rules of procedure created by members of society across many generations. Individuals contribute but little to the creation and alteration of the rules; even those in high places find their freedom of action circumscribed by the privileges and obligations of their position or office.

The sum of the rules that code the essential operations of society constitutes a basic information pool, possessed by all members of society collectively. The collective information pool is equivalent to the culture of a society, when culture is broadly defined to include the characteristic features of all human behaviors, and not merely the "high culture" of science, art, and religion.

If society's information pool (its broadly defined "culture") is up to date and operational, the production and consumption systems function adequately to maintain society in its milieu. All the essential flows are replenished; all the basic subsystems are repaired or reproduced. Society is sustainable; its people are in harmony with each other and in balance with their environment. On the other hand, society is critically unstable—and on the point of a bifurcation—if it is unable to replenish the flows and repair or reproduce its subsystems. In this unstable and nonsustainable

112 THE REALMS OF EVOLUTION

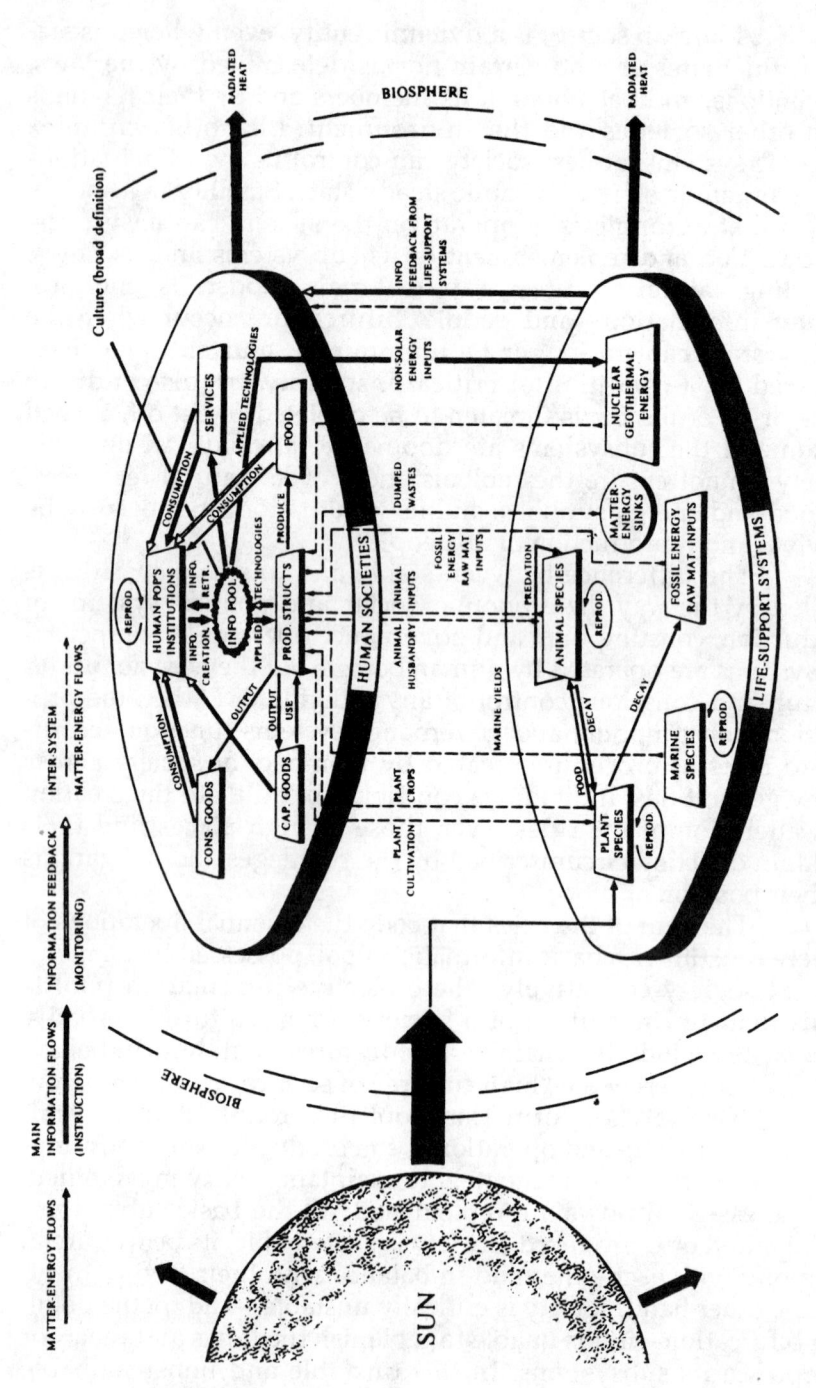

Figure 5.2. Auto- and cross-catalytic cycles in sociocultural systems

Figure 5.2 maps the main flows in the biosphere, integrating the ecosystemic catalytic cycles of Figure 2.4 with the additional flows and cycles introduced by human societies.

Neglecting a large number of subsidiary flows and cycles, the basic processes are easy to grasp. Energy to drive the entire system comes from the sun (a comparatively small amount is now also derived from sources of nonsolar origin, such as nuclear and geothermal energies). Plants convert solar energy into biomass and provide food for animals. The organic wastes and decaying bodies of animals are, in turn, food for plants. Plants and animals as well as all marine species reproduce themselves partly through natural processes in their ecosystems, partly through human intervention (plant cultivation, animal husbandry, maricultures—although the latter are as yet insignificant on the global scale). The natural life-support systems also include fossil energies (oil, coal, natural gas), the result of fossilized plant decay over eons of time, and matter and energy sinks on land, in the water, and in the atmosphere.

On the level of human society, we note human populations constantly creating and conserving information (know-how, skills, blueprints, laws, norms, and so on). The sum of such information represents the currently operative information pool of society. It is directly accessed by people in libraries, data banks, the press, and many other ways. Much of the collective information pool is applied to produce the goods and services needed or demanded by human populations. The production structures include industrial as well as agricultural production. They operate on the basis of information, but use energy and matter inputs from the life-support systems (fossil and renewable or nuclear energies, plants, animals, marine species, and so on). The production systems also replenish the renewable part of these inputs, but if they take out more than they allow to grow back, even renewable stocks diminish over time.

People in groups and institutions monitor most of these flows: there is an information feedback loop from many parts of human activities as well as from the life-support systems. The information thus fed back is assessed by people and enters the common information pool: the broadly defined collective culture of a society. Destabilizations can occur if there is a major culture lag: that is, if information is incomplete or deficient, and if it is not acted upon in the production systems. The production systems themselves produce food and manufactured goods for human consumption, as well as capital goods for use within the production systems themselves. Services, an increasingly important part of the economic life of societies, use relatively little energy and materials; their main commodity is information. Information processed in the service sector is likewise consumed by the human populations, and is in part cycled back into the information pool.

Last but not least, it is important to note that all these complex flows and cycles (which in reality are far more complex than this figure can represent) serve to maintain human beings, enabling them to survive and to reproduce. In addition, they are also self-serving on the level of human society: the flows and cycles likewise maintain the institutions created by human populations. As long as all cycles are sustainable, the system can buffer out minor perturbations and continue to operate. But when some cycles are interrupted or collapse, major fluctuations shake the entire system, which must then transform, or perish.

condition the members of society must update and transform their culture, that is, their collective information pool. They either manage this feat and attain a new and functional mode of social, economic, and political organization, or their society lapses into anarchy and may dissolve in chaos, a ready prey to more stable and powerful adversaries.

The dynamics of social evolution concern the progressive yet discontinuous development of society's collective information pool. The processes of history act on actual societies but they select cultural information pools, much as the processes of biological evolution act on actual organisms but select genetic information pools. This, then, is the deeper meaning of the term *cultural evolution*.

When destabilized by uncontrollable fluctuations, a society does not suffer extinction but is absorbed in other societies—or it transforms and renews itself. In any event, it undergoes a major phase change equivalent to a bifurcation. The historical record provides ample evidence of the occurrence of such bifurcations. As we shall now illustrate, at least three major varieties appear to have occurred with significant frequency: T-bifurcations due to the destabilizing effect of technologies; C-bifurcations induced by conflicts, and E-bifurcations due to economic and related social problems.

THE VARIETIES OF SOCIAL BIFURCATIONS: A HANDFUL OF CASE STUDIES

1 . T-Bifurcations

The relentless spread of modern technologies creates imbalances within contemporary societies. Established values are uprooted, new expectations are encouraged, and, as they are frequently unfulfilled, frustration and discontent are fueled. Technologies spreading from a handful of industrialized and postindustrialized countries to the far corners of the world change people's lives. They trigger social change, but often at high human cost.

The Hoplite Revolution

Technological innovations have triggered social transformations in the past; only the dimensions of the challenges posed by current technologies are new and not the challenges themselves. Humankind's first experiment with democracy is a case in point. Known as the hoplite revolution, it was a successful response to the challenge of a little-known—but highly significant—technological innovation.

The innovation concerned warfare. Around 700 B.C., iron became common and inexpensive enough for average citizens to possess. It was used to make offensive weapons as well as protective armor. This seemingly minor innovation was enough to effect some truly basic changes in Hellenic societies. In the Mycenaean age, warfare was the privileged domain of aristocrats, who could afford to purchase horses and bronze armor; chariots and cavalry played the dominant role. The privileged few led the battles and reaped the rewards of victory. But when iron technologies became widely available, military prowess came to the foot soldier. Against an infantry clad in armor and wielding iron thrusting spears, cavalry was almost always defeated. The "hoplite phalanx," as it was called (*hoplite* comes from the Greek *hoplon*, meaning "tool"), relied on compactness rather than on individual action as it broke through the enemy ranks to throw their line into chaos.

Hoplites became the most effective fighters in the Mediterranean world. Within fifty years they eroded the power of the aristocrats and took over the business of governance as well as that of warfare. Hoplites were not whipped into battle, as Herodotus writes that the Persians were. They fought on the basis of consensus developed through the logic of argument. Consequently the power that came to the citizens through the spread of iron technology triggered a reorganization that, in the Hellenic society of the epoch, was expressed in the principle of *isonomia*—equal participation in government. The social transformation that was thus created laid the foundations of Western democracy: the advent of the hoplites is aptly called a revolution.

The Second Industrial Revolution

Unlike the revolution of the hoplites, the revolution triggered by the new technologies of information, communication, and automation is not history: it is living reality. We cannot as yet foresee its outcome, but can already assess its nature and dimensions.

Probably more than 90 percent of all the scientific research and development ever undertaken in history has occurred in the period since the beginning of World War II. Scientific R&D, which has accelerated continuously in most countries, gave societies the power of the atom as well as the power of information, communication, and automation. It has put unprecedented power in the hands of those who control the new technologies and now threatens the lives of those who are exposed to their unreflective use—or their purposive misuse. Nuclear technologies, for example, can be used in explosives, and they can destroy the human population of the planet twenty times over. They can also render the biosphere unfit for life higher than insects and grass. But if nuclear technologies are used for peaceful purposes, deserts can be irrigated, fresh water can be made universally available, machinery basic for industry and agriculture can be powered, and life and prosperity can be assured not only for six, but for ten or more billion people. The planet's carrying capacity can be enlarged and the demands of an increasingly large and increasingly prosperous population can be satisfied.

However, such a "nuclear paradise" would call for a major reorganization of contemporary societies. It is doubtful whether people today have the vision and the will to carry it out. Einstein may have been right when he said that the unleashed power of the atom has changed everything save our modes of thinking. If in the nuclear age humanity is to prosper—and even if it is merely to survive—current modes of thinking and behavior would have to change radically. Our most basic forms of social, economic, and political organization will have to be reformed. It is by no means clear whether the twentieth-century system of nation-states could survive into the twenty-first century: the system may be too prone to conflict and too adverse to cooperation to permit the spread of nuclear technologies for peaceful uses alone. It is also questionable whether the current economic and financial system could survive—its concentration of economic power in the hands of national governments and giant transnational corporations may be inimical to the development and application of the technologies needed not only to support the wealthy and the educated, but also to feed, house, employ, and educate the poor and the illiterate.

The problems are not eliminated if, instead of mainly nuclear, a mix of new energy technologies is used. Indeed, a proper mix of renewable and nonrenewable, nuclear and solar-based technologies may prove to be the optimum solution to the energy problems of our economies; but a mix of energy sources would require further societal adjustments. Solar technologies harness the energy of the sun, and, as such energy reaches the earth in a diffuse form, the applications of the technologies tend to be decentralized and small-scale. To use solar power as a main energy source for metro-industrial complexes would call for covering large surface areas with solar cells, solar towers, windmills, or other technological devices. Since this is both impracticable and environmentally hazardous, solar technologies would best be used in small, decentralized economic units. Thus attention would have to shift to many as yet underdeveloped rural areas. Farmers and villagers would have to be given the opportunity to develop the technologies in light of their own needs and preferences. This would call for major shifts in the allocation of priorities between capitals and hinterlands, shifts that many governments advocate in principle but few are willing to carry out in practice.

Mastering the information technologies of the second industrial revolution calls for similarly profound societal changes. The gap between advanced information societies and the majority of developing countries in the Third World is growing with astonishing speed. It polarizes the international economic system. Multinationals alone are not likely to be effective in bridging the gap, as their primary goals are profit and growth and not the service of socioeconomic equality and justice. Yet if current trends are not turned around, massive worldwide unemployment will result as automated control technologies—the already famed "robotics"—replace unskilled and semiskilled human labor in country after country and field after field.

The latest information technologies deplace jobs even in the advanced information sector. Unlike previous information technologies that permitted people to do things they earlier could not do, the new technologies also do things that people rely on being *able* to do. Modern computers reduce the number of draftsmen and engineers required to see projects through to completion. They reduce the number of clerks and secretaries needed in business and administration. They deplace a number of professionals, such as lawyers and medical doctors. Computerized "expert systems" diagnose illnesses, identify errors, give financial and legal advice, prepare taxes, and do an increasing number of jobs hitherto entrusted to people. Indeed, employment in the information sector of advanced societies has all but stopped growing in recent years, and may actually begin to shrink.

The technologies of the second industrial revolution are truly revolutionary. Not only are they major advances over previous technologies, but they also require major transformations in the societies that make use of them. A postindustrial society based on abundant and dense energy sources and on information, robotics, and automation reaches new heights in autonomy: it progressively detaches itself from geographic constraints. It generates its own energies and produces its own raw materials with less and less dependence on the endowments of its milieu. This could lead to a brusque rupture of geocultural ties which evolved over centuries; customs, forms of organization, and systems of beliefs mediating between society and environment could suddenly become irrelevant, or actually obstructive. The traditional aptitudes of the work force, strongly tied to social, cultural, and historical customs and traditions, could become all but useless. Skills essential in a preindustrial community may be marginally useful in an industrial society, and could prove to be entirely superfluous in a postindustrial one.

The new technologies make distinctions between city and countryside almost as irrelevant as geographic location. Space and time constraints are all but eliminated by telecommunications and high speed transportation. There is no further need for urban megacomplexes; high-tech producers and consumers can just as well live in decentralized suburban or rural settings.

The new technologies are not passive "menus" from which individuals, firms, and societies can pick and choose according to their capacities and appetites. The technologies come in mutually supporting clusters, engulfing the user in a vast array of procedures that he or she may be hard put to master and manage. For example, CIM—computer-integrated manufacturing—is a rational innovation for a wide variety of manufacturers. It integrates the entire manufacturing process from the original product concept to the procurement of components, their assembly and inventory management, and their transport and marketing. CIM calls for CAD (computer-aided design) and for OA (office automation). Economic logic, and the survival imperative of firms in high-pressure competition, dictates that these clusters of technologies be used together. Success seems to crown the efforts of those who master the full range of the available technologies and do not hesitate to innovate wherever and whenever innovation promises benefits.

But in the late twentieth century, success of this kind may be a short-term affair, and its pursuit a double-edged sword. The innovators may find that in creating a postindustrial enterprise in an industrial (not to mention a preindustrial) society, they have bit off more than they could chew. Or they may succeed in mastering their own procedures, only to find that they are overproducing goods that are in limited demand. Traditional production systems geared to supply standardized mass-produced goods are not easy to transform into footloose, decentralized, and personalized goods-producing enterprises—especially if the transformation requires firing much of the work force. Consumers in today's societies may become impoverished and alienated by depersonalized robots, even if the robots churn out personalized goods and custom-tailored services. Mistrust combined with lack of buying power could thwart the shining rationality of managers keen on technological innovation.

The technologies of the second industrial revolution are not simple extensions of the classical industrial technologies with improvements in quality and quantity, speed and reliability. The new technologies are fundamentally different, and they call for fundamentally new mind-sets in individuals and new forms of social and economic (and hence also political) organization in society.

The new technologies are clearly aligned with long-term trends in the evolution of human societies: they conduce to high levels of autonomy and greater complexity, and access to abundant, and in part also highly dense, energy sources. But to achieve the bifurcation that still separates us from fully evolved postindustrial societies, we would need to transform our thinking as well as our social structures. Ignoring "soft" sociocultural factors would soon turn them into surprisingly hard and obstructive realities.

2. C-Bifurcations

History provides a rich terrain for the exploration of conflict-induced bifurcations in the development of societies. Unlike technological and economic bifurcations, these dramatic—and often also traumatic—events are normally perceived as revolutions in the traditional sense of the term.

C-bifurcations (the political variety of revolutions) do not necessarily mean a complete break with the past; historians have often pointed to elements of continuity. In the case of the French Revolution of 1789, Tocqueville found that it accomplished many goals pursued by French kings since Louis XI, and Lord Acton announced that it brought the political structure of French society into line with long-established traditions in its social structure. But historians have also discovered that revolutions change the established institutions of society, adapting them to the forces and factors—technological, economic, or political—that brought down the previous regime. Taking an optimistic evolutionary view, Henri Sée sought to demonstrate that revolutions speed up rates of growth, purge ossified institutions, and establish spiritual goals that stimulate further development. Later historians, such as Chalmers Johnson, offered detailed analyses of modern revolutions conceived as the adjustment of the state apparatus to societal dysfunctions. Although there is still no agreement among contemporary historians as to what causes revolutions, whether they follow a uniform pattern, and what their consequences are, most scholars would now agree that revolutions, despite certain continuities, also produce discontinuities that are rapid and significant. Many would also claim that the changes can be understood as reorientations of society to radically altered conditions. Revolutions seem to be functional from the viewpoint of creating institutions, behaviors, and values that respond to the threats and challenges that destabilized the "ancient regime."

The world wars of the twentieth century produced the widest variety of conflict-triggered revolutions in recent history. Among them, two stand out in virtue of their worldwide impact: the transformation of tsarist Russia into the Soviet Union, and the transformation of the Weimar Republic into the Third Reich.[1]

Bifurcation in Tsarist Russia

The rise of the Communist regime in Russia, like that of the Nazi regime in Germany, was triggered by the giant wave of political, social, and economic instability created by World War I. This was a "total" war for Europe: millions of casualties were suffered; immense debts were accumulated; social relationships were drastically altered; and economic, financial, and basic moral resources were used up. By the end the experience had exhausted almost all combatants, and states lacking the resilience to cope were driven to the point of breakdown.

Although the revolutionary changes that took place in Russia in 1917 and in Germany fifteen years later were triggered by the instabilities created by World War I, they had deep historical roots. Russia's failure to modernize its social, economic, and political structures during the nineteenth century left it unstable and rigid; it was readily destabilized by a major war. Indeed, failure in the 1904 war with Japan had already led to a nearly successful revolt in 1905. Attempts within Russia to accelerate reforms, such as those of Stolypin, were frustrated by conservative political forces. But economic growth, industrialization, and social unrest proceeded apace and created growing tensions. The result was a government that alienated its middle-class supporters while driving the peasants and workers toward ever more violent acts that antagonized the foes of the regime. By 1914 the nation was so divided internally that the outbreak of war merely delayed the revolution by momentarily redirecting crucial social and political energies.

Nicholas II's inability to govern created a deep crisis as the Russian armies suffered a series of humiliating defeats. The tsar's decision to take personal command only led to his being personally blamed for the failures. Corruption, aggravated by the ascension of Rasputin over the tsarina—because of his strange power to control the hemophilia of the crown prince—meant that the state bureaucracy was even less able to supply desperately needed arms, clothing, and food. By February 1917, food riots had broken out, the Russian economic and political system had collapsed, and the Romanov dynasty had been toppled.

The speed and thoroughness of the collapse of the tsarist regime surprised even Sukhanov and other who had devoted their lives to the revolutionary overthrow of the government. Even more surprising was the final outcome of the revolution. Stalin and the Petrograd leadership had supported the provisional government for months. *Pravda* actually printed articles opposing plans by Lenin and Trotsky to "telescope" stages by seizing power in November. At the critical moment Lenin himself despaired of success, and only Trotsky was able to dissuade him from fleeing to Finland. But the situation at home became unsustainable and ripe for change. The war had precipitated the collapse of the tsarist regime, and Allied demands that Russia honor its treaties and continue fighting made it impossible for the provisional government to solve the basic problems that had caused the February revolution. The Bolsheviks succeeded in stepping in at the crucial moment. Even though they were but a relatively minor political movement—a "peripheral isolate"—they could seize power in the power vacuum and chaos of the day. Divided among themselves and scattered from Geneva to Brooklyn, their ideas were often contradictory and their strategies unclear. But Lenin came out with his "April Theses," and his ideas and strategy became increasingly attractive in the deepening crisis. His slogan "Bread, Land and Peace," promised to meet the most basic needs of the masses. It levered the Bolsheviks into position as leaders of the urban workers, whose prewar violence had been contained only briefly by patriotic appeals. Internal class antagonisms were greatly exacerbated by oppressive policies, poor military performance, and gross corruption. Increasing radicalism among urban groups as well as among peasants and soldiers, combined with Lenin's political and oratorical skills, Trotsky's genius for military organization, and the efficiency of the revitalized Bolshevik apparatus tipped the scales by November: the Bolsheviks seized the reigns of power.

Bifurcation in the Weimar Republic

Germany's transformation from the Weimar Republic to the Third Reich was less telescoped in time than the events in Russia but was hardly less dramatic. Having entered World War I as the strongest power on the Continent, Germany's defeat created a major social trauma among its people. Scapegoats had to be found to explain how a magnificent army, countless acts of personal courage, and extensive national sacrifices culminated in humiliating losses of territory, sovereignty, foreign colonies, a powerful fleet, and great industrial might. Mutual recriminations between conservatives and socialists, republicans and monarchists, and between various segments of society and Jews led to serious internal fragmentation. The rapid introduction of modern technology into Wilhelmian Germany had already destabilized the traditional institutions and created deep and widespread alienation. Compounded by defeat and the humiliation of the Treaty of Versailles, social institutions collapsed and revolts became a frequent occurrence. Street fighting was rampant, "political" murders like those of Erzberger and Rathenau became common, and civil war seemed imminent.

War debts, the penury inflicted by the Treaty of Versailles, foreign interventions, and a devastating inflation kept the German economy in crisis until the Young Plan was instituted in 1932. By then the Depression (precipitated by the failure of an Austrian bank, which broke the essential cycle of payments) had driven unemployment to all-time highs. The divided Reichstag was unable to maintain a majority; after 1930, government was by presidential decree. Yet the president, Field Marshal Hindenburg, was too aged and fragile to exercise the needed leadership. He was easily manipulated by intriguing army officers like Schleicher and Papen.

The competition between these two officers was a major factor in Hitler's rise to power in January 1933. In times of relative normalcy, Adolf Hitler, the Austrian corporal who had failed as an architect and whose accent was so alien that Prussians ridiculed it, would have been an unlikely choice in a society as sophisticated as the Weimar Republic. His party platform was so contradictory that he himself promised not to implement it, and his previous attempts to seize power (during the 1923 financial crisis) had ended in imprisonment. Yet, in the chaos of a destabilized society, Hitler was able to mobilize political support. A month after announcing that there was "not the slightest chance" of the Nazis coming to power, and three days after asserting that Hitler could never take over the government, Hindenburg himself named Hitler chancellor. Papen had persuaded Hindenburg to "tame" Hitler by giving him responsibilities and letting him demonstrate his incompetence. But Hitler was by this time the head of the largest party in the Reichstag, and his coalition gave him a real though slight majority. His personal appeal to the masses gave the disintegrating society a sense of unity. In the wake of the Reichstag fire—whose cause remains obscure to this day—he used the power of party and of propaganda to build up his charisma and make himself invincible.

Hitler's policies were not without appeal to the remaining wealthy classes since he promised to respect property, prevent strikes, and fight communism. His appeal for the masses was reinforced by aligning himself with an anti-Semitic tradition that traced its roots to the Middle Ages. Extolling the virtues of the Aryan race and the destiny of the German state, and promising to revenge lost wealth and prestige by bringing Germany to the pinnacle of world power, Hitler's atavistic policies and personality had the crucial effect: it overcame the frustration and alienation of the postwar decade and rallied society to a new cause—however human and violent that cause may appear in historical perspective.

The rise of the Bolsheviks in Russia and the rise of Hitler in Germany are dissimilar in many particulars: no attempt is made here to draw moral and ideological parallels. Yet, like more recent transformations in China, Cuba, Iran, and hosts of other Third World countries, they exhibit the basic dynamic of conflict-triggered C-bifurcations.

3. E-Bifurcations

Bifurcations triggered by economic and social crises are of particular relevance to the future: they may become comparatively widespread in coming years. While such E-bifurcations occurred sporadically in history, a massive wave of economic-instability-induced societal changes may come about in the last decade of this century because of a fateful combination of population growth and resource deprivation. The problems created by these factors are further exacerbated by the increasingly unbearable burden placed on today's economies by military expenditures.

The spread of modern technologies plays havoc with the demographic balance of traditional societies. New health-care technologies and improved diets decrease infant mortality and extend life expectancy in country after country. But they do not directly impact on traditional desires for large families: the demographic curve shows a significant delay in the decrease of the birthrate.

Many societies in the developing world find themselves unable to keep up with the demands of their people. Under the weight of a rapidly growing human population with exponentially increasing demands, the supply of several kinds of basic resources will inevitably run low. Such resource squeezes could provoke severe economic recession and accompanying social and political instabilities. While the threshold of resource-deprivation-based instability is likely to vary from society to society, depending on the specific nature of their systems of production and consumption, and on the level of efficiency and discipline of the people, it seems clear that beyond certain limits social structures will become critically unstable.

Looking at the resource picture on the level of the globe as a whole, there is cause for concern. World population has grown explosively during the twentieth century and will continue to grow rapidly in coming decades. Even a reduction in fertility rates cannot significantly slow the still ongoing explosion of the world population: some 35 percent of the world's living people are children under fifteen years of age. Even if those who now enter the age of fertility have fewer children than their parents, the number of people who will be *added* to the human population in the last quarter of this century would still be greater than the entire world population at the beginning of this century (2.05 billion people, compared to 1.6 billion). The over six billion humans who will populate this earth by the year 2000 will also demand far more resources per capita than any generation has ever demanded before. It seems likely that humanity will consume more resources during the second half of the twentieth century than it did during the previous three hundred centuries of its existence as a dominant species.

The fact that most of the increase in the world population takes place in the poor developing regions of the world does not improve the picture. Tradition-bound poor countries would have to develop new and optimally efficient technologies to access, store, and utilize a wide variety of energy and material resources—and do so without destroying the integrity of their life-support systems. This is feasible in principle, but it does call for a major reorganization of the dominant systems of production, consumption, and governance.

CONCLUSIONS

There is pattern in history, as there is pattern in the evolution of order and complexity in all realms of nature. Even if outcomes are never predetermined, it is not accidental that, given sufficient time, life should appear, and then society. It is logical, in terms of the dynamics of evolution, that on each level evolving systems should acquire greater complexity and retain more of the energy fluxes in which they are immersed. The emergence of mammals with endothermy is just as much in the logic of evolution as the emergence of technological societies. As the one is a third-state *organic* system capable of storing and using increasing amounts of increasingly dense energies on the level of individual organisms, so the other is a third-state *supraorganic* system with equivalent capabilities on the level of groups and populations of organisms.

The adoption of increasingly energy-intensive technologies with improved (though not necessarily optimal) efficiency sets forth the evolutionary trend toward growing dynamism and autonomy in nature. Just as organic species evolve toward the use of greater densities of a wider variety of free-energy sources in their environment, so human societies develop to access, store, and use in greater densities larger quantities of free energy through the ongoing improvement of their technologies. As a consequence societies, the same as natural systems, tend to grow larger in size, develop more intricate relations among their diverse components, and create more massive and flexible modes of interaction among them.

The evolutionary hypothesis makes sense. History can be conceived as the evolution of systems in the third state on organizational levels typical of societies of living beings. The detailed exploration of this hypothesis should be high on the list of research priorities of social and natural scientists alike.

Notes

1. A number of social scientists have tried to identify the nature of bonds in modern societies. Emile Durkheim suggested "social facts" and "solidarity" as explanatory concepts; Weber pioneered attempts to pin down the binding force of institutions in the social roles played by individuals, while Kurt Lewin produced actual measurements of such forces in terms of quasi-physical factors such as vectors, boundaries, and barriers. Likewise the experiments of Solomon Asch and Stanley Milgram contribute to the clarification of the nature of social bonds by identifying and measuring the operation of social constraints in terms of conformity and authority.

2. For example, a comprehensive report to the National Science Foundation defined technology as "any tool or technique, any physical equipment or method of doing or making, by which human capability is extended." L. G. Tornatzky et al., *The Process of Technological Innovation: Reviewing the Literature* (Washington, D.C.: National Science Foundation. 1983).

3. Technology as agent of social change leads neither to optimum energy efficiency nor to maximum energy consumption. In some sense consumption is the inverse of efficiency: the more energy one consumes to perform a given task, the less efficient one is. Of course, the energy requirement of the tasks performed in societies tends to grow as well. But as more energy-intensive tasks are adopted, energy efficiency can actually decrease. This was the case in the shift from preindustrial to industrial society. Energy consumption per capita went up, parallel with energy waste. By contrast the shift from industrial to postindustrial society decreases energy consumption per capita by increasing the efficiency with which energy is employed.

    Taken separately, neither energy efficiency nor gross energy consumption is the critical measure. That measure is the overall increase in the dynamism of society, i.e., the amount of free energy actually used to maintain the social system together with its population, its implements, and its infrastructure. If the accessed free-energy flux is abundant, consumption tends to go up and efficiency tends to go down. If the flux is dense rather than abundant, efficiency increases. While in history, energy consumption and efficiency fluctuate, often inversely, the energy used to maintain the social system shows an overall increase. There is a growth in the dynamism of society and therewith in its autonomy in the natural environment.

6

The Evolution of Mind

The phenomenon of mind is perhaps the most remarkable of all the phenomena of the lived and experienced world. Its explanation belongs to the grand tradition of philosophy—to the perennial "great questions" that each generation of thinkers answers anew . . . or despairs of answering at all.

A general theory of evolution could, as we shall see, shed some meaningful light on this complex and elusive issue.

THE ORIGINS OF MENTAL FACULTIES

Mental faculties, of which the most basic is a form of intelligence, are not unique to man: other species have developed forms of them, and would have developed them further, had they the need and the opportunity to do so. Whales and dolphins have intelligence, but they live in an aquatic environment that is both more stable and more friendly than life on land. Consequently sea mammals had no need to evolve their intelligence in the way that land-living mammals did. The latter needed an intelligence capable of manipulating the immediate environment, since in terrestrial settings the availability and retention of water, the ongoing procurement of free energy, and the maintenance of a constant temperature are essential to insuring the integrity of complex biochemical reactions.

The origins of mental faculties in humans can be traced to the time when the early hominids diverged from the tree-dwelling apes. They then found themselves in a perilous situation. The savannahs were already populated with meat-eating animals, most of them stronger and faster than they. The shelter of the trees was gone, and in its place the adventurous hominids had only one substitute: their newly freed forelimbs. These were no longer needed to hold on to the branches of trees and could thus be put to other uses. Most probably, evolving arms were used for self-defense with stones and sticks, and to transport infants when hominid bands migrated with their herds of animals on Africa's developing grasslands. Our ancestors learned to depend for their survival on a high level of bodily control, tactile sensitivity, and manual dexterity.

As forelimbs transformed into dexterous arms and hands, jaws were no longer required for defense. There was also no selection pressure for large canine teeth, sectorial premolars, and a capacious jaw to accommodate them. The pressure was for a larger brain, capable of a more precise coordination of bodily functions and a more selective manipulation of the immediate surroundings.

A larger brain permitted the development of abilities that were advantageous not only on the level of the individual, but also on the level of the group. The early humans learned to cooperate in performing critical tasks of survival. As individuals with a superior ability to communicate among themselves diffused, the genetically based sign language of apes transformed into the system of shared symbols of human languages. With language, social behavior was freed from the rigidity of genetic programming and became adaptable to changing circumstances. In the evolving neocortex of humans, capacities for manual dexterity and tool use were joined with capacities for analytic perception, language, and socialization.

In human bands and tribes, the edge of survival in competition with other species was assured both by greater individual dexterity and a more analytic perception, and by the enhanced ability of individuals to work together in pursuing shared goals and projects. Then, as now, many tasks that are essential for daily existence could be better achieved through teamwork than by individual effort. Effective teamwork—in hunting game, tending plants and animals, as well as caring for home and family—called for detailed and reliable communication among the members of a group, and this was beyond the limits of the sign-language used by animals. Through long processes of trial and error, our ancestors mutated the ability to develop the kind of language that uses consensually defined symbolism in addition to (and ever more in preference) to direct signs.

The use of a symbolic language enabled our ancestors to identify recurring objects and events, and to identify themselves as well. This laid the basis for a higher type of consciousness: a consciousness not only of the world around the perceiving individual, but of that individual in the world. The phenomenon we know as mind entered the scene of evolution on this planet.

THE NATURE AND CONTINUED EVOLUTION OF MIND

In the evolutionary perspective (unlike in the creationist view), mind is not a static phenomenon, given once and for all. It is a dynamic faculty that emerges in the processes of evolution, and it is subject to evolution in itself. While the evolution of mind is similar in many respects to the evolution of third-state systems in general, mind is not a third-state system, nor any other system of physical components and physical, chemical, or biological processes. Rather, mind is a specific faculty manifested by a complex third-state system: the human being.

Mind, in the form of a stream of conscious experience that accompanies us from birth to death, is more than the mere perception of objects, people, and events: it is also their cognition and recognition, together with the thoughts, feelings, motivations and intuitions that accompany our perceptions. Human conscious experience is qualitatively different from the objects and events that appear in that stream when it is focussed on some aspect of the world around the perceiver. Thus, as philosophers have often pointed out, we cannot investigate the human mind with the methods used to investigate the human brain, or indeed any matter-energy system in the universe. Thoughts, images, feelings, and sensations are "private." None of us has direct access to the mind of anyone else—not even of his closest friend or relative. Mind can only be investigated through introspection.

The private nature of mental phenomena and their sui generis method of investigation need not embroil us in the egocentric predicament—in the solipsist affirmation that all we can know is our own mind. We can connect our private introspections with public knowledge, building the immediately known facts and events of our consciousness into the conceptually constructed world of empirical science. Rather than assume that "my mind" is all that I can know of the world, we can assume that a scientific reconstruction of the nature of reality has validity—that it can give us as good and certain knowledge as it is humanly possible to get. While such knowledge does not have eternal validity, it does represent our most qualified guesses concerning the nature of the reality that lies somewhere beyond our mind, in public time and space. Thus we can dare to assume that the universe exists and evolves; even that in the universe systems in the third state exist and evolve. If we do make these assumptions we come to the logical conclusion that we ourselves are such a system. We can then pose the more reasonable and manageable question: what is the significance of the sequence of perceptions, thoughts, and feelings that we ourselves experience as mind?

The bundle of sensations that makes up the experience of mind can be most logically and parsimoniously explained as the internal description of the relations between an organism (body-cum-brain) and the environment. If mind is a function and a manifestation of a complex nonequilibrium system, it is not necessarily restricted to the human species. More likely it is a highly sophisticated variant of the kind of undifferentiated "feel" of the milieu that less complex systems have on the multicellular, cellular, or even subcellular levels of organization, corresponding to their sensitivity (on lower levels merely irritability) vis-à-vis their environment. Their sensitivity, in turn, is a function of the specific nature of their survival requirements. As evolution brings forth more and more complex systems, further and further from equilibrium, the requirements of survival become ever more stringent. They call for increasingly sophisticated capacities for distinguishing the relevant features of the environment, and for searching out the appropriate sources of free energies. In the physiology of evolving biological species, the nervous system evolves from primitive ganglia into a complex brain.

If mind is indeed an internal description of the relations between organism and environment, then, parallel with the evolution of the brain, mind evolves from an undifferentiated sea of primitive sensations to a locus of differentiated perceptions. As the brain becomes the organ for seeking out dense and highly specific forms of free-energy fluxes with ever more reliance on information and learning, the mind, being the internal read-out of the operations of the brain, becomes the means for orienting the organism with ever-greater refinement toward ever-greater autonomy in its surroundings.

The human mind, however, is not just the subjective side of a two-sided survival mechanism. The mind, as introspection reveals, is also the seat of abstract thought, feeling, imagination, and value. I not only sense the world, I also interpret my sensations. Like presumably all human beings, I have consciousness. I am aware of having sensations and, on successively higher levels of abstraction, I am aware of being aware of having sensations. Ultimately I, like other members of the human species, learn to abstract from immediate sensations in ways that lesser species cannot, and can come to deal with pure forms of thought. These include scientific and mathematical concepts, aesthetic constructions, and the abstract meaning of words and concepts. Consciousness is not a mysterious transcendental trait: it is the capacity for internally describing the internal description of the perceived and conceived environment.[1]

Neurophysiologists find that consciousness in human subjects is associated with the functioning of the frontal lobes of the neocortex. As these regions were the last to evolve—perhaps only about 200,000 to 250,000 years ago—consciousness as we know it may have been experienced only by a handful of hominid species, and for the last 30,000 years only by *sapiens*. This, however, does not mean that other species, and in fact all systems in the third state, would not have internal descriptions of their relations with the environment. The lack of a large neocortex means only that they cannot analyze these descriptions on levels of precision and abstraction that human beings can.

A system capable of both sensing its environment and analyzing the description of its sensations is a highly sophisticated entity: it knows its milieu, and knows that it is knowing it. It can identify sets of sensory perceptions with objects and events, and establish relations and connections among them. It can come to believe that it knows not only its own stream of sensations but the external world itself. Such knowledge is, of course, hypothetical. It is testable, however, since it is prone to error. The possibility of error is important: it is the key not only to the validity of knowledge, but also to its improvement. Error is the price paid for learning. Were living systems to persist in basic errors, they would soon be weeded out by natural selection. In humans such selection would operate first of all in the competition for essential sources of free energy (air, water, and food) and then in the equally crucial competition for information (knowledge, skills, social orientation, etc.).

Species that survive do not persist in basic errors but learn to correct them. The new solutions may likewise contain errors that come to light subsequently. Thus an endless vista is opened for learning. And human learning, being empirical rather than genetic, is relatively rapid: it takes place in the lifetime of individuals, not of species.

Learning leads to the emergence of differences between members of the same species and creates true individuals. The brain is known to develop much as the embryo develops: by progressive differentiation among groups of cells. The process is not determined directly by the genes but indirectly by the combined effect of genes and signals from cell groups that activate genes: it is epigenetic rather than genetic. Just as identical twins are never truly identical, already because of the differentiation of the epigenetic development of their physiology, so the brains of two individuals, even if they could be exposed to the same set of stimuli, could never be identical owing to the fact that different neuronal groups would be activated by the stimuli.

Individually differentiated brains are the external, observer oriented dimension of the individually differentiated minds presented "internally" to each human being. As the brain maps incoming stimuli in a variety of ways by sorting them according to various, uniquely differentiated "maps," the mind perceives and recognizes the environing world in various, individually unique ways. As the environment changes, so do the multiple categories of perception in the mind and the multiple mappings in the brain. Unique individuals cope with the world in unique ways. Thus, like all systems in the third state, humans have a basic "divergence property." Because they are highly evolved systems with complex and differentiated brains, the divergent pathways of their development lead to individuals who are highly differentiated in their psychology, and not only in their physiology.

But we should not separate the evolution of mind from the evolution of society: the two take place simultaneously and in close relation. Because the human brain is capable of multiple mappings, and the mind of empirical learning, human societies are capable of rapid evolution. They move from survival-oriented hunting-gathering tribes and primitive agrarian communities to so-called civilized societies where culture plays a dominant role and survival is assured by an ever smaller segment of the population in an ever shorter period of time. The individual becomes increasingly free to learn: to explore the world, liberated from the survival chores of finding food, clothing, and shelter.

The reduction of the segment of the population that takes care of basic survival needs is a historical phenomenon. As the social system becomes more dynamic, while the biological requirements of its human population remain constant, more and more of the dynamic capability of society is channeled into activities that have little to do with physical survival. The process has been known to social scientists for centuries. Already in 1691 the British economist William Petty wrote that "as Trade and the Curious Arts increase; so the Trade of Husbandry will decrease." In 1938 Australian economist Colin Clark divided the economy into three sectors, agriculture, manufacturing, and services, and noted that each of the sectors flowered in turn. This process holds good today. In the United States the service sector had outgrown the previously dominant manufacturing sector already in 1960. Japan, the society that is perhaps the furthest along its historical trajectory, no longer has a full-time agricultural work force. The fastest growing sector in most of the European economies is the one concerned with information processing and related services.

In its irreversible development technology not merely saves labor; it detaches it from the survival chores and functions of people. It sifts the bulk of the work force from hunting, fishing, and gathering food, and later from growing it, to building abodes, drawing boundaries, administering lands, producing and storing goods, performing services, and finally to intellectual, esthetic, and spiritual pursuits.

As the needs of survival become filled in society, people experience other needs, including the "higher" needs for esthetic environment, intellectual satisfaction, and ultimate meaning. And as people pursue the satisfaction of these needs, society acquires a cultural superstructure, connected with, but no longer subservient to, the feedbacks and catalytic cycles that maintain a social system in its environment.

The human mind, triggering the evolution of ever more technologically advanced societies, frees itself from the sphere of survival and creates the sphere of culture. If it now succeeds in mastering the increasingly automated technologies of postindustrial societies, it will have the opportunity to evolve still more varied cultural pursuits, still further removed from the chores of physical survival.

Mind does not evolve in a vacuum, nor indeed does it evolve in the isolated individual. Mind and culture coevolve with technology and society: a technologically sophisticated society is at the same time the product of the evolving mind, and its generating environment.

From vulnerability to changes in the environment—to basic irritability and responsiveness to change—to differentiated perception and selective behavior that not merely responds but creates and initiates, mind evolves as the subjective dimension of the objective survive functions of complex nonequilibrium systems. Freed by technological innovation from time-consuming survival chores, and driven by the need to find outlets for its creative energies and impulses, the learning mind creates the world of culture. The human mind evolves as technological society evolves, and they both evolve as systems in the third state do in nature: toward greater complexity and autonomy in interaction with the environment.

A CONCLUDING NOTE

This review of the great realms of evolution—the physical realm of the evolution of matter, the biological realm of the evolution of life, the realm of history that encompasses the evolution of human society, and the realm of psychology concerned with the development and operations of the mind—explored the empirical application of the general theory outlined in Part One. Before closing these explorations, we should recall that it is not possible to decide the truth of this theory—or of any conceptual scheme for that matter—by comparing it with pristine reality. There is no immaculate perception; we do not see reality except through the spectacles of a theory. But not all spectacles allow coherent vision, not all theories make sense of the strands of order that we intuitively believe interconnect the myriad phenomena encountered in experience. The best we can do to prove the truth and validity of our theory is to show that, when we use it as a spectacle to view reality, reality appears coherent and meaningful.

The requirement for coherence and meaning underlies all efforts at our understanding of the world; more than this no concept or theory can achieve. If a general evolution theory discloses coherence and meaning in reality, it can rightfully claim to be true, at least until shown to be otherwise. In any case, its further testing and exploration will be well worth our time and effort.

Notes

1. In terms of the technologies of artificial intelligence, consciousness (and by this I mean the self-reflective, self-analytical capacity of the mind and not the basic phenomenon of feeling and sensation) is the functional equivalent of a subsystem designed to monitor the functioning of another system. The basic data of the subsystem are the signals passing to and fro among the elements of the host system. The monitor analyzes these data according to algorithms wired into its circuits. There may be subsidiary monitors that analyze data flows in the primary monitor. If a sequence consisting of a large number of monitors were to be built into an artificial system, the system would model the functional aspects of human consciousness.

From the
Midnight Notebook

We are not strangers in an alien, mechanistic universe—the general theory of evolution reintegrates us with nature. Our distant ancestors considered themselves an organic part of all that exists, expressing their belief through myth and confirming it through magic. Our not-so-distant forebears thought they knew better and put man above nature, and indeed beyond it. Now we are back in the embrace, indeed in the womb, of a creative universe, a universe capable of bringing forth life, and even consciousness.

Modern science, it is said, has dethroned man, removing him from center stage and banishing him to a small planet in an indifferent solar system near the edge of a galaxy. But a general theory of evolution enthrones man; it looks not at *where* we are but at *what* we are. We are one of the most remarkable expressions of nature's thrust toward order, structure and system, a thrust toward the hazardous realms of nonequilibrium where survival calls for sophistication and skill . . . and ultimately for mind and for intelligence.

Our new self-respect need not be exaggerated. We are surely among the evolutionary avant-garde, but we are not the highest-level system, even in our planetary environment. We are not the top of the evolutionary hierarchy, we are a part of larger systems. The global ecology—the quasi-living and breathing Gaia system—is the largest of the systems in our immediate environment. And within it there are innumerable large-scale systems: ecologies, made up of plants, insects, and marine and terrestrial creatures of all kinds, as well as human societies, made up of ourselves and our fellow humans in a maze of intersecting and shifting relations. Recognizing our "parthood" does not detract from our status as the most complex creature in this, and perhaps in any, region of the universe. And it does remove the anthropocentric blindfold that prevents us from seeing a great opportunity—and a grave threat. The opportunity is to consciously and purposively evolve society, the carrier of human evolution in our age. The threat is that the rapid evolution of increasingly complex structures in society may stifle the individual, locking him into relations that, while eminently functional in assuring dynamic stability in society, may be inhuman and constraining for its members.

To the best of our knowledge there is but one system in all the universe of which the parts are conscious while the whole is not, and that is human society. What is real wisdom and good ethics under these circumstances? If there should be a conflict between the welfare of the part and the evolution of the whole, should we not champion the former rather than promote the latter? Evolutionary ethics is not necessarily the ethics of evolution—of evolution *sub specie eternitatis*. It could be the ethics of human freedom and creativity: the ethics of nurturing a faculty in the part even if it means putting a brake on the advance of the whole. Surely, if the part is one of the most noteworthy embodiments of the creativity of the cosmos, it has a right to some measure of self-interest.

Evolution, in cutting across time-honored but now antiquated disciplinary boundaries, contradicts common sense as much as modern physics ever did. Not only is the solid three-dimensional table on which I write a whirling dance of electrons and a vibrating symphony of molecules, but the electrons, atoms and molecules of the universe are not fundamentally different from the tree outside the window and the dog by the fireplace. And none of them is fundamentally different from me and other human beings—nor from the technological society in which we live. Never has the classical French witticism, *plus ça change plus c'est la même chose*, had as much meaning as it does today. To perceive only the diversity is obsolete common sense, and to perceive only the unity is nonsense. To perceive the unity within the diversity may be the genuine sense; a sense disclosed by the evolutionary vision.

For humanity, evolution is history. But history has been the prey of chance choices among unknown or dimly perceived alternatives. History—the chance-ridden evolution of human societies—has been recorded, dissected and debated *ex post facto*: it has not been studied and steered *ante factum*. Yet it can be so studied and steered: as evolution becomes history, it can become conscious. As Jonas Salk put it: conscious evolution can emerge from the evolution of consciousness—and from the consciousness of evolution.

In a system such as contemporary society, evolution is always a promise and devolution always a threat. No system comes with a guarantee of ongoing evolution. The challenge is real. To ignore it is to play dice with all we have. To accept it is not to play God—it is to become an instrument of whatever divine purpose infuses this universe.

Hamlet's question is not whether to evolve or not to evolve, but whether to evolve with distinction or devolve to extinction. For evolution is liberal but not permissive: it does not let us stay fixed in our tracks, nor does it let us reverse our steps. But it does allow us to go forward as best or as poorly as we can. *"Know Thyself"*, challenged the oracle at ancient Delphi. *"Know how you evolve"*, is the challenge we face today.

Evolution is not like the advance of the hands of a clock, moving relentlessly forward in fulfillment of preordained laws. Evolution is not fate; to evolve is not to meet one's destiny. We are not puppets on a cosmic stage, acting out a play whose script has been written and sealed. We are actors with will and purpose, and the power to vary the script—at least our part in it. We can vary it wisely or foolishly, to the extent of our wisdom . . . or our folly.

This is an exciting age, perhaps the most exciting in the history of humanity. We live at the precise moment when we are simultaneously becoming aware of the processes that evolve our societies, and are acquiring a mastery over the technologies that determine how they evolve. We live at the conjunction of knowledge and power. Whether we also live at the moment of genuine wisdom remains to be seen.

". . . 'tis the worst of times, 'tis the best of times, 'tis the age of foolishness, 'tis the spring of hope, 'tis the winter of despair, we have nothing before us, we have everything before us." Thus wrote Charles Dickens in 19th century England, in the turmoil of the first industrial revolution. He could have been writing of the world of the late 20th century, in the grips of the second (or post-) industrial revolution. For that much bandied about but little known process, the progress of civilization, is not without its risks. It was with good reason that Alfred North Whitehead noted that the greatest advances in civilization have all but wrecked the societies that produced them. The technologies of information, communication and automation, the torchbearers of societal evolution in our age, may yet wreck some contemporary societies—and the people who live in them.

Pessimism is premature; optimism is naive. Voltaire was right: the optimist believes that we live in the best of all possible worlds, and the pessimist is afraid that this is true. Neither does anything about it—the one because nothing *needs* to be done about it, and the other because nothing *can* be done. The evolutionary realist knows that he can shape the world in which he lives—and so he acts.

In the beginning there was chaos, instability, inflation and radiation. Within an almost infinitesimal fraction of a second the first microparticles evolved. After half a million years stable atoms—matter in the non-ionized state—appeared. Within five billion years the galaxies began to take shape and then the stars. In the last three billion years life has emerged on this earth. For the last hundred thousand years, hominid creatures with conscious minds roamed this planet. And for the past several thousand years we sapiens have wondered: Where have we come from—and where are we going? Today, some 15 billion years since the origins of the universe, we may be close to an answer.

Appendix:
The Evolution of Science

Does GET apply to science—and thus to itself?

The question is intriguing. Would the general theory of evolution describe its own genesis?

The answer is not evident. The general theory of evolution refers to dynamic matter-energy systems in the empirical world; but scientific theories themselves are conceptual systems, mental constructs created by scientists. It is true, on the other hand, that scientific theories arise through the interaction and intercommunication of scientists with each other and with "nature;" that is, that they emerge in the context of the interaction of some highly specific matter-energy systems with each other and with their environment.

A general evolution theory would reflect and describe its own genesis if scientific development would indeed manifest enduring patterns, and if these patterns would prove to be isomorphic with those traced by the emergence of matter-energy systems in the empirical world. To show whether or not this is the case would require detailed studies in the history, philosophy, and sociology of science. This is beyond the scope of this Appendix. But we can make a modest beginning by noting some salient points.

Since the publication in 1962 of Thomas Kuhn's *The Structure of Scientific Revolutions* it is seldom contested that scientific development follows definite patterns—regardless of whether these are exactly such as Kuhn suggested. The course of scientific innovation is not entirely random but possesses its own internal logic. Karl Popper, Imre Lakatos, Stephen Toulmin, Mario Bunge, and many other philosophers of science produced scores of learned treatises to specify this logic; they showed how scientists and scientific communities actually operate and found that maverick theorist Paul Feyerabend is off the mark when he claims that in science "anything goes."

The concept of "revolution" adds a saltatory element to the logic of scientific development. Although revolutions are sometimes contested in regard to Kuhn's original formulation (some philosophers, including Popper and Toulmin, argue that they occur more or less at all times although not in equal measure), it is generally acknowledged that scientific progress is nonlinear, marked by periodic bursts of innovation interspersing phases of relatively routine research. During periods of "normal science," fundamental tenets are not questioned; theories inspired by the underlying paradigms seem to accord with reality. Scientists are mainly conservative: they content themselves with devising additional tests for existing theories through better instrumentation and new observational techniques, and with extending the theories to as yet uninvestigated phenomena. As a result, in normal science periods, theories cover and account for more and more phenomena without fundamentally altering their conceptual structure and their underlying assumptions.

However, theories in the empirical sciences are strongly coupled to their natural environment—the specific aspect of reality that they describe and explain—and are exposed to "perturbations" from it. Environmental perturbations come in the form of new evidence (observation or instrument reading) that fails to conform to expectations and predictions flowing from the theories. Such anomalies can accumulate and cast doubt on the soundness of the theoretical core. When theories accumulate a critical mass of anomalies, their status changes: they no longer command the allegiance of the representative majority of scientists in the pertinent discipline. Although some scientists remain conservative and fight rear-guard action to preserve the anomaly-ridden theory, others turn revolutionary: they begin to question its most basic assumptions. Only by such revolutionary questioning could the centuries-old dominion of Newton's physics come to an end around the turn of the century: it was Einstein's fundamental insight that the mechanistic world view underlying classical physics was not to be rescued by adding equations to equations—"epicycles" to "epicycles"—to explain the anomalies. Instead, he produced the theories of special and general relativity: a fundamentally new conception of the physical universe.

Once the great revolution in physics ran its course and belief in the eternal verity of scientific truths was broken, improved techniques of observation and experimentation, together with more refined analytical techniques produced an entire series of revolutions—in biology and biochemistry, in psychology, and not the least in the social sciences. Though not flattening into a continuous straight line, the path of scientific development became marked by more frequent leaps from one paradigm to another.

The transformations denoted by "revolutions" or "paradigm shifts" not only change the basic constructs of science; they also simplify them. One of the main features of newborn theoretical conceptions is their simplicity compared with the maze of explanations which burden the theories they replace. The comparative simplicity of a construct is not to be confused with its level of comprehensibility. The concepts that make up the new theories are seldom if ever easy to grasp. As a rule they are abstract, removed from the plane of observation and common sense. For example, the formula $E = mc^2$ is simpler than the Lorentz transformation formulas by which physicists tried to salvage the Newtonian equations in view of the anomalies of black-body radiation and the constant velocity of light regardless of the motion of the observer, but it is also more abstract than anything produced by Newtonian physics. The invariance of the Einsteinian interval in four-dimensional space-time is likewise a simple proposition, but its meaning is almost mysterious when contrasted with the Euclidean three dimensional space and the equitably flowing one-dimensional time that frame Newton's equations of motion.

Revolutionary innovations in science order and simplify formerly cluttered and chaotic conceptual systems, but do so at the cost of greater abstraction. There was progressive simplification through abstraction in field after field of modern science; we need merely to compare the basically simple structure of Darwinian biology with the complex classification scheme of Linnaeus; the streamlined models of modern behaviorism with the esoteric theories of nineteenth-century depth psychologists, or the key conceptions of general equilibrium theory with the speculations of pre-Keynesian economists.

Yet, except for gains in simplicity and explanatory power, the outcome of scientific revolutions cannot be predicted. There is nothing predetermined about the contents of the theory that will surface following the demise of its predecessor; the only requirement is that it explain the facts explained by the previous theory plus those that are anomalous for it, and that it do so with optimum economy and simplicity. Theories consisting of a number of different concepts and based on a number of different assumptions could conceivably satisfy these requirements. According to the Mach-Duhem-Poincaré theorem, an indefinite number of theories can fit any given set of observations.

Whether or not the number of candidate theories is large, it is certainly more than one. This condition provides effective choice for science in the revolutionary periods of theory innovation. Which of the several possible candidate theories is actually selected depends in large measure on such chance factors as personal insight and the moment when the hypothesis is actually presented. Timing is more important than is usually conceded: it is worth recalling that Darwin was sufficiently uneasy about the implications of his theory in nineteenth-century England that he hesitated some twenty years before presenting his ideas on the origin of species, and that Einstein succeeded in gaining adherents for his special relativity concepts only when physicists were becoming desperate for lack of a consistent and manageable theoretical framework.

It is entirely possible (though not predictable) that a general theory of evolution will provide the basis for the next paradigm of contemporary science. The expectation is not without foundation. There is, first of all, a distinct need for a new paradigm. The social sciences are struggling with a set of competing conceptions, and the sciences of complexity introduce ideas that are new and disturbing in view of their origin in the natural sciences. Advances in nonequilibrium thermodynamics, in cosmology, in quantum biology, and in consciousness research create upheavals in their respective fields and lead many researchers to believe that a new scientific revolution may be just around the corner. At the same time, society at large places increasing blame on science for the havoc wrought by haphazard uses of technology and is frustrated for not receiving clear-cut answers from the scientific community.

The chaos needed to open a period of scientific revolution may be already at hand. The search for a new paradigm is likely to be weighed in favor of candidates that can integrate recent breakthroughs in important scientific fields in a coherent general conception. The successful candidate will be simple yet abstract, general yet adequate to describe and explain a wide variety of phenomena. Its coming will probably fit the overall pattern of scientific development. This pattern has been incisively described by F.S.C. Northrop. Science, in his view, progresses from an early "natural history" phase where emphasis is on fieldwork and there tend to be as many independent propositions and assumptions as there are direct observations, toward a mature, deductively formulated stage where the already gathered observations are analyzed and described in terms of the smallest possible number of primitive concepts and elementary propositions. As Northrop noted, the primitive terms of the mature theories are significantly more general than the concepts of the natural history phase. This is clearly illustrated in the evolution of the law of gravity. Galileo's law applies to all falling bodies on the surface of the earth; Newton's law applies to all bodies whether on earth or elsewhere in the universe; and Einstein's law applies to all moving bodies in space-time, even under acceleration and at velocities approaching the speed of light.

Unlike the laws of physics and of other specific scientific fields, the concepts and postulates of a general evolutionary theory apply to a wide variety of disciplines in the empirical sciences. The interdisciplinary—or rather *trans*disciplinary—scope of GET is an important qualification in the search for a new scientific paradigm. As the Japanese scientist Kinhide Mushakoji points out, no phenomenon can be adequately known apart from the overall system of natural—and sometimes also social—realities in which it is embedded. Although individual researchers cannot cover all aspects of nature and society within the scope of their special competence, they *can* frame their research problems and assess their findings in light of general theories of wide multidisciplinary scope. A general evolution theory allows specialized investigators to divest themselves of the blindfolds that normally accompany specialty vision and permits them to situate their particular segment of the empirical world within the relevant wider context.

GET is not unique in crossing disciplinary boundaries and connecting diverse fields of empirical investigation: almost all the sciences of complexity are multidisciplinary in scope. Ross Ashby, one of the pioneers of these sciences, once remarked that the circulation of money through banks in an economy and the circulation of blood through the kidneys in the body are essentially the same process, however odd it may seem to collapse physiology and finance. Indeed, Ashby's self-stabilizing and self-organizing model known as the "homeostat" applies equally to artificial servomechanisms and to the human brain. Norbert Wiener's cybernetics describes processes of control in animals as well as in machines; Ludwig von Bertalanffy's general system theory applies to the behavior of systems regardless of the nature of their parts or components; and the laws of contemporary thermodynamics apply to all energy-processing systems whether they are natural or artificial, living or nonliving.

The fact that a theory arches over several fields does not make it any the less valid than theories that stay within single disciplines. Nature does not observe disciplinary boundaries even if scientists do. But even scientists do so less and less today, as the relative isolation of the specialized communities of scientists gives way to intercommunication and the comparison of suppositions, methods and results. It is not surprising to encounter a growing number of general theories that embrace and integrate diverse disciplinary structures. Current trends in theory formulation conduce toward higher levels of abstraction and generality combined with greater simplicity and economy not only within specific fields, but also across them; hence they lead ultimately to general theories of transdisciplinary scope. Attempts to advance general theories of evolution are merely the latest, and in respect to generality the most advanced, manifestation of this trend.

The trend, noted here, toward higher levels of abstraction and generality with greater simplicity and economy across disciplinary boundaries appears to be a secular one, enduring throughout the age of science. As long as scientists and scientific communities embrace the values they embraced since Galileo and Newton, they will be oriented toward general theories of the evolutionary variety. Indeed, general evolution theories may constitute the basic ideal of science in the long term.

The fact that the development of scientific theories is patterned, saltatory, unpredictable, and self-simplifying, and lifts the conceptual apparatus of science to ever-greater heights of abstraction and generality (analogously to the domains of nonequilibrium reached by evolving matter-energy systems) suggests that the development of complex interactive systems, whether they pursue their careers in the real world of matter-energy flows or in the conceptual realm of ideas, follows basically similar dynamic patterns. Very likely these patterns indicate the ultimate limits and possibilities of the evolution of complex systems in a changing environment.[1]

The development of science is itself an evolutionary process. A mature theory would reflect and explain itself; a fully developed science of evolution would also describe the evolution of science.

Notes

1. It is possible, and may prove to be fruitful, to investigate these patterns in depth and show that the similarities between the evolution of matter-energy systems in the empirical world and the evolution of conceptual systems in science are more than mere analogies. A model of scientific development could be proposed where theories are dynamic systems of constructs in interaction with their environment, consisting of other theories and theory developments, and the direct or indirect observation of "nature." Paradigms would represent the set of attractors that define the state of the construct system. During standard phases of "normal science," the attractors are stable (periodic or static) and regulate the development of the system of constructs in a deterministic manner. In periods of major innovation ("revolutions"), the attractors turn chaotic, and there are catastrophic bifurcations in the course of which attractors that define the outgoing paradigm disappear and those that define the incoming paradigm appear "out of the blue" (through the genius of innovators and the serendipity of the spread of their ideas). If such models prove to be reasonably accurate and dependable representations of the dynamics of theory elaboration and innovation, the development of science would appear as a genuine evolutionary process, and the applications of GET would be extended to the history, sociology, and philosophy of science.

Bibliography

NONEQUILIBRIUM THERMODYNAMICS AND DYNAMIC SYSTEMS THEORY

Abraham, Ralph and C. Shaw, *Dynamics: The Geometry of Behavior*. 4 Vols. Santa Cruz: Aerial Press, 1984-1988.
Eigen, Manfred and P. Schuster, *The Hypercycle: A Principle of Self-Organization*. New York: Springer, 1979.
Epstein, Irving, K. Kustin, P. de Kepper and M. Orban, Oscillating Chemical Reactions, in *Scientific American*, March 1983.
Glansdorff, P. and I. Prigogine, *Thermodynamic Theory Structure. Stability and Fluctuations*. New York: Wiley Interscience, 1971.
Katchalsky, Aharon, Biological Flow Structures and their Relations to Chemodiffusional Coupling. *Neuroscience Research Progress Bulletin* 9, 1971.
Katchalsky, Aharon and P.F.Curran, *Nonequilibrium Thermodynamics in Biophysics*, Cambridge, Mass. MIT Press, 1965.
Kaufmann, Stuart, *The Origins of Order: Self-Organization and Selection in Evolution*. Oxford: Oxford University Press, 1993.

Maturana, Humberto R, and Francisco Varela, *Autopoietic Systems*. Report BCL 9.4. Urbana: Biological Computer Laboratory, University of Illinois, 1975.
Nicolis, G. and I.Prigogine, *Self-Organization in Non-Equilibrium Systems*. New York: Wiley Interscience, 1977.
Onsager, L., Reciprocal Relations in Irreversible Processes. *Physiological Review*, 38, 1931.
Prigogine, Ilya, *Etude Thermodynamique des Phenomenes Irreversibles*. Liege: Desoer, 1947.
_____, *Thermodynamics of Irreversible Processes*, 3rd. ed. New York: Wiley Interscience, New York, 1967.
_____, and I. Stengers, *Order Out of Chaos* (La Nouvelle Alliance). New York: Bantam, 1984.
The Science and Praxis of Complexity. Tokyo: The United Nations University, 1985.
Thom, René, *Structural Stability and Morphogenesis*. Reading, Mass: Benjamin, 1972.
Varela, Francisco, H.R. Maturana and R. Uribe, Autopoiesis: The Organization of Living Systems, Its Characterization and a Model. *Bio-Systems* 5, 1974.
Zeeman, Christopher, *Catastrophe Theory*. Reading, Mass: Benjamin, 1977.

GENERAL AND EVOLUTIONARY SYSTEMS THEORY

Bertalanffy, Ludwig von: *General System Theory*. New York: Braziller, 1968.
Bohm, David, *Wholeness and the Implicate Order*. London: Routledge & Kegan Paul, 1980.
_____, and J.B. Hiley, *The Undivided Universe*. London: Routledge, 1993.
Boulding, Kenneth E., *Ecodynamics, a New Theory of Societal Evolution*. Beverly Hills and London: Sage, 1978.
Capra, Fitzjot, *The Web of Life*. New York: Avon Books, 1996.
Ceruti, Mauro, *Constraints and Possibilities: The Evolution of Knowledge and the Knowledge of Evolution*. New York: Gordon & Breach, 1994.
Combs, Allan, *Cooperation: Beyond the Age of Competition*. New York: Gordon & Breach, 1992.
Corning, Peter A., *The Synergism Hypothesis, A Theory of Progressive Evolution*. New York: McGraw-Hill, 1983.
Csányi, Vilmos, *General Theory of Evolution*. Studia Biologica Hungarica 18, Budapest: Akademiai Kiado, 1982.
_____, *Evolutionary Systems and Society: A General Theory*. Durham and London: Duke University Press, 1989.
Goerner, S.J., *Chaos and the Evolving Ecological Universe*. New York: Gordon & Breach, 1994.

Goertzel, Ben, *The Evolving Mind*. New York: Gordon & Breach, 1993.
Gray, William and Nicolas Rizzo, (eds.), *Unity Through Diversity* (2 Vols.). New York: Gordon & Breach, 1973
Haken, Hermann, *Synergetics*. New York, Springer, 1978.
Haken, Hermann (ed.), *Dynamics of Synergetic Systems*. New York: Springer, 1980.
Harris, Errol E., *Cosmos and Anthropos*. New York: Humanities Press, 1991.
Hayek, Friedrich A. von: Kinds of Order in Society. *Studies in Social Theory* No. 5. Menlo Park, Institute for Humane Studies, 1975.
Hoyle, Fred, *The Intelligent Universe*. London: Michael Joseph, 1983.
Jantsch, Erich, *Design for Evolution*. New York: Braziller, 1975.
_____, *The Self-Organizing Universe*. Oxford: Pergamon Press, 1980.
_____, and Conrad H. Waddington, (eds.): *Evolution and Consciousness*. Reading, Mass: Addison-Wesley, 1976.
Laszlo, Ervin, *Introduction to Systems Philosophy*. New York: Gordon & Breach, 1972 (reprinted 1984); 2nd ed. Harper Torchbooks, 1973.
_____, (ed.), *The New Evolutionary Paradigm*. New York: Gordon & Breach, 1991.
_____, *The Age of Bifurcation: Understanding the Changing World*. New York: Gordon & Breach, 1991.
_____, I. Masulli, R. Artigiani and V. Csanyi, *The Evolution of Cognitive Maps: New Paradigms for the Twenty-First Century*. New York: Gordon & Breach, 1993.
_____, *The Creative Cosmos: A Unified Theory of Matter, Life, and Mind*. Edinburgh, Floris, 1993.
_____, *The Interconnected Universe: Conceptual Foundations of Transdisciplinary Unified Theory*. London and Singapore: World Scientific, 1995.
_____, *The Whispering Pond*. London and New York: Elements Books, 1996.
Margenau, Henry (ed.), *Integrative Principles of Modern Thought*. New York: Gordon & Breach, 1972.
Masulli, Ignazio, *Nature and History: The Evolutionary Approach for Social Scientists*. New York: Gordon & Breach, 1990.
Pattee, Howard (ed.), *Hierarchy Theory: The Challenge of Complex Systems*. New York: Braziller, 1983.
Salk, Jonas, *The Anatomy of Reality*. New York: Columbia University Press, 1984.
_____, *The Survival of the Wisest*. New York: Harper & Row, 1973.
Weiss, Paul A., et al. *Hierarchically Organized Systems in Theory and Practice*. New York: Hafner, 1971.

BIOLOGICAL EVOLUTION

Ager, D.V., *The Nature of the Stratigraphic Record.* New York, Wiley, 1973.
Carson, H.L., "The genetics of speciation at the diploid level," *American Naturalist*, Vol. 109, 1975.
Corliss, John B., "On the creation of living cells in submarine hot spring flow reactors." *Proceedings, Firth ISSOL Meeting and Eighth International Conference on the Origin of Life*, Berkeley, California, 1986.
Denton, Michael, *Evolution: Theory in Crisis.* London: Burnett Books, 1986.
Eldredge, Niles, *Time Frames: The Rethinking of Darwinian Evolution and the Theory of Punctuated Equilibria.* New York: Simon and Schuster, 1985.
_____, *Unfinished Synthesis: Biological Hierarchies and Modern Evolutionary Thought.* Oxford: Oxford University Press, 1985.
_____, Stability, Diversity and Speciation in Paleozoic Epeiric Seas, *Journal of Paleontology*, Vol. 48, 1974.
Eldredge, Niles and Stephen J. Gould, Punctuated Equilibria: an Alternative to Phylogenetic Gradualism, in Schopf, ed.: *Models in Paleobiology.* San Francisco: Freeman, Cooper, 1972.
_____, Punctuated Equilibria: "The tempo and mode of evolution reconsidered," *Paleobiology.* Vol. 3, 1977.
Frohlich, H. ed., *Biological Coherence and Response to External Stimuli.* Heidelberg: Springer, 1988.
Goodwin, Brian, "Development and evolution," *Journal of Theoretical Biology*, 97, 1982.
_____, "Organisms and minds as organic forms," *Leonardo*, 22, 1, 1989.
Gould, Stephen J., *Ever Since Darwin.* New York: Norton, 1980.
_____, "Darwinism and the expansion of evolutionary theory," Science, Vol. 210, 1980.
_____, *Ontogeny and Philogeny.* Cambridge, Mass.: Harvard, 1977.
Grimes, W. and K.J. Aufderheide, *Cellular Aspects of Pattern Formation: The Problem of Assembly.* New York and Basel: Karger, 1991.
Guttmann, F. and H. Keyzer, eds. *Modern Bioelectrochemistry.* New York: Plenum, 1986.
Hall, Barry G. "Evolution on a petri dish," *Evolutionary Biology*, 15, 1982.
Ho, Mae-Wo, "On not holding nature still: evolution by process, not by consequences," *Evolutionary Processes and Metaphors*, M.-W. Ho and S.W. Fox, eds., London: Wiley, 1988.
_____, *The Rainbow and the Worm.* Singapore and London: World Scientific, 1995.

Holling, C.S., "Resilience and stability of ecosystems," in Jantsch and Waddington, eds.: *Evolution and Consciousness*. Reading, Mass: Addison-Wesley, 1976.
Lotka, A.J., *Elements of Physical Biology*. Baltimore: Williams & Witkins, 1925.
Morowitz, H.J., *Energy Flow in Biology*. New York: Academic Press, 1968.
Oparin, A.I., *The Origin of Life*. New York: MacMillan, 1938.
Pribram, Karl, *Brain and Perception: Holonomy and Structure in Figural Processing*. Hillsdale, NJ: Lawrence Erlbaum, 1991.
Sheldrake, Rupert, *A New Science Of Life*. London: Blond & Briggs, 1981.
_____, *The Presence of the Past*. New York: Times Books, 1988.
_____, *The Rebirth of Nature*. New York: Bantam, 1991.
Stanley, S.M., A Theory of Evolution Above the Species Level. *Proceedings of the National Academy of Sciences* USA 72, 1975.
Volterra, V., *Leçons sur la théorie mathematique de la lutte pour la vie*. Paris: Gauthier-Villars, 1931.

COSMIC EVOLUTION

Chaisson, Eric J., *Cosmic Dawn: The Origin of Matter and Life*. Boston: Atlantic, Little, Brown, 1981.
_____ *The Life Era: The Role of Change in the Natural Universe*. Boston: Atlantic Monthly Press, 1986.
_____ *Universe: An Evolutionary Approach to Astronomy*. Englewood Cliffs: Prentice-Hall, 1988.
Gunzig, E., J. Géhéniau and I. Prigogine, "Entropy and cosmology," *Nature*, 330, December 1987.
Hawking, S.W., G.W. Gibbons and S.T.C. Siklos, eds., *The Very Early Universe*. Cambridge: Cambridge University Press, 1983.
Hoyle, Fred, G. Burbidge and J.V. Narlikar, "A quasi-steady state cosmology model with creation of matter," *The Astrophysical Journal*, 410, June 20, 1993.
Kafatos, M. ed., *Bell's Theorem, Quantum Theory and Conceptions of the Universe*. New York: Kluwer Academic Publishers, New York, 1989.
Prigogine, Ilya, J. Géhéniau, E. Gunzig, and P. Nardone, "Thermodynamics of cosmological matter creation," *Proceedings of the National Academy of Sciences USA*, 85, 1988.
Trefil, J.S.: *The Moment of Creation: Big Bang Physics* New York: Scribner, 1983.

Index

A
anomalies, 142
attractors, 42-46
attractors, chaotic, 44-45
autopoietic systems (autopoietic systems theory), 3, 22, 40-41, 110

B
Bénard cells, 32, 33
bifurcation, 45-46, 93
bifurcation in history, 108-111
Big Bang, 65, 67
bonding energy 25, 27
brain, 131, 132

C
C-bifurcations, 119-124
catalytic cycles, 33-34, 37, 38, 46, 50, 85
catastophe theory, 22, 42-46
celestial mechanics, 14
chance, 38, 83

change, 2, 5, 16, 37, 50
change, irreversible, 2, 5
change, nonlinear, 40
chaos (chaos theory), 3, 42-46
classical mechanics, 16
complex systems, 21, 28, 37, 71, 147
conscious experience, 129
continuum of evolution, 25-30
convergence, 34-37, 46, 89, 90, 98
Copernican theory, 14-15
cultural evolution, 114
cybernetics, 22

D
Darwinian biology, 16
demographic balance, 123-124
destabilization, 38, 50, 114
determinism, 22-23, 96
development, 105
DNA, 88
double helix, 88

Index

E
E-Bifurcations, 123-124
élan vital, 18
energy efficiency, 107
equilibrium, 17, 18, 22, 24, 88, 94
equilibrium, chemical, 32
error, 132
eternal becoming, 13
evidence, 142
evolution, incremental, 84
evolution, laws of, 60
evolution, logic of, 30
evolutionary axioms of society, 97-98
evolutionary paradigm, 11-19, 23
evolutionary paradigm, antecedents of, 12-15
evolutionary philosophies, 2
evolutionary trajectory, 38
evolutionary vision, 12
evolvere, 2

F
flow reactors, 81-82
fluctuation, 38, 46, 110
fossil record, 84
free energy, 25, 30-34, 125, 130, 131
free energy flow, 33
free energy flux density, 31, 46, 50

G
galaxy formation, 70
general evolution theory (GET), 2, 3, 6, 17, 21-52, 59, 63, 127, 135, 141, 144, 145
general system theory, 2, 21
generalists, 85
genome, 87-89
genotype, 87, 88
grand unified theory (GUT), 64-65
gravity, 145
Great Chain of Being, 14

H
hierarchy, 27-28, 37, 50
history, 96, 99, 105, 108-111, 125
hominids, 91, 128
Homo, 91-94, 101
Hoplite revolution, 114-115
hypercycles 34, 46, 89, 90

I
indeterminacy, 37, 46
industrial revolution, first, 104
industrial revolution, second, 115-119
inflationary scenarios, 65-66
information pool, 111
instability, 37-38
invariance, 2, 6
invariant pattern, 5
Ionian natural philosophers, 13
irreversibility, 67, 105
irreversibility, thermodynamic, 144

L
language, 128-129
learning, 132
life, evolution of (biological evolution, 79-94
life, evolutionary dynamics of, 83-91
life, origins of, 80-83
living systems, 60

M
matter, evolution of, 63-75
matter-energy systems, 60
matter-energy systems, synthesis of, 67-72
meaning, 3, 4, 133, 135
mental faculties, origins of 127-129
Middle Ages, 103
mind, evolution of, 125-135
mutation, 87-89

Index

N
natural philosophy, 15
Neolithic, 102
noumenon, 15

O
observations, 144
order, 5, 6, 7, 32
organizational levels, 25, 27-30, 30-37, 40, 46, 60, 89, 98, 125
origins of life, 33

P
perceptions, 131, 134
phenotype, 87
predictability, 38, 93
progress, 99, 105
punctuated equilibria, 84, 85, 108

R
rediation-to-matter energy transfer, 72-75
revolution, Franch, 119
revolution, scientific, 142-145
revolutions, 119-120

S
sapiens, emergence of, 91-94
science, 4, 12, 15, 21, 59-60, 141-147
science, method of, 4, 7
scientific law, 22-23
sensations, 130
social bifurcations, varieties of, 114-124
social darwinism, 2
social evolution, direction of, 99-108
society, evolution of (social evolution), 95-125
sociocultural systems, 60

specialists, 85
speciation, 83-87, 110
split between natural sciences and humanities, 96
steady-state, 38
stellar evolution, 70
suprasystem/subsystem, 27-28
synthetic theory, 83
systems in the third state, 23-25, 30-31, 38-40, 73, 97, 98, 110, 125, 132, 134

T
T-bifurcations, 114-119
technologies, information 117-118
technologies, nuclear, 115-116
technologies, solar, 116
technology, 100-108, 115, 119, 123, 125, 133
thermodynamics, classical, 16, 17
thermodynamics, irreversible (nonequilibrium), 3, 18, 22
thermodynamics, second law of, 17, 18, 24
time, 16, 17, 19, 72-73, 91
time, two arrows of, 15-19
transformation, 37-38
truth, 135
Tsarist Russia, 120-121
turbulence, 44
two cultures, 15

U
universe, origins of, 63-66

W
Weimar Republic, 122-123
world population, 124